Superconductivity:
An Introduction

Superconductivity: An Introduction

Martha Butler

Larsen & Keller
www.larsen-keller.com

Superconductivity: An Introduction
Martha Butler
ISBN: 978-1-64172-687-0 (Hardback)

Larsen & Keller

Published by Larsen and Keller Education,
5 Penn Plaza,
19th Floor,
New York, NY 10001, USA

Cataloging-in-Publication Data

Superconductivity : an introduction / Martha Butler.
    p. cm.
Includes bibliographical references and index.
ISBN 978-1-64172-687-0
1. Superconductivity. 2. Superfluidity. 3. Electric conductivity. I. Butler, Martha.
QC611.92 .S86 2022
537.623--dc23

For more information regarding Larsen and Keller Education and its products, please visit the publisher's website www.larsen-keller.com

Table of Contents

Preface

The set of certain physical properties which are seen in specific materials is known as superconductivity. Due to these physical properties, the electrical resistance vanishes and the magnetic flux fields are expelled. The materials which show these properties are known as superconductors. The resistance in normal metallic conductors decreases gradually when the temperature is lowered even close to absolute zero. In superconductors, there is a distinctive critical temperature, below which the resistance suddenly drops to zero. Therefore, a loop of superconducting wire can keep on conducting electric current indefinitely, without any power source. This book traces the progress of this field and highlights some of its key concepts. Most of the topics introduced in it cover new techniques and the applications of superconductivity. The book is appropriate for students seeking detailed information in this area as well as for experts.

To facilitate a deeper understanding of the contents of this book a short introduction of every chapter is written below:

Chapter 1- The property of certain materials wherein electrical resistance tends to become zero is called superconductivity. It is divided into low-temperature superconductivity and high-temperature superconductivity. This chapter has been carefully written to provide an easy understanding of the varied facts of superconductivity.

Chapter 2- Superconductivity is a vast subject that comprises of several models and theories. The London equations, Ginzburg-Landau theory, Bardeen-Cooper-Schrieffer theory, Bean critical state model in superconductivity are some examples. This chapter discusses in detail these models and theories related to superconductivity.

Chapter 3- Any material which conducts electric current without any resistance is referred to as a superconductor. There are different materials that exhibit superconductivity such as mercury, lead, niobium-titanium, magnesium diboride, etc. These are further classified as type 1 and 2 superconductors. The topics elaborated in this chapter will help in gaining a better perspective about superconductors.

Chapter 4- There are various special material-based superconductors. These include A-15 compounds, CeRu2, pyrochlore oxides, rutheno-cuprates, magnetic superconductors or chevrel phases, heavy fermion superconductors, oxide superconductors without copper, etc. This chapter closely examines these material-based superconductors to provide an extensive understanding of the subject.

Chapter 5- Superconductors have a wide range of applications in the fields of power industries, automobiles, medicine, etc. They are also used in maglev trains, magnetic resonance imaging and nuclear magnetic resonance machines. These diverse applications of superconductors have been thoroughly discussed in this chapter.

I would like to share the credit of this book with my editorial team who worked tirelessly on this book. I owe the completion of this book to the never-ending support of my family, who supported me throughout the project.

Martha Butler

Understanding Superconductivity

The property of certain materials wherein electrical resistance tends to become zero is called superconductivity. It is divided into low-temperature superconductivity and high-temperature superconductivity. This chapter has been carefully written to provide an easy understanding of the varied facts of superconductivity.

Superconductivity is a complete disappearance of electrical resistance in various solids when they are cooled below a characteristic temperature. This temperature, called the transition temperature, varies for different materials but generally is below 20 K (−253 °C).

The use of superconductors in magnets is limited by the fact that strong magnetic fields above a certain critical value, depending upon the material, cause a superconductor to revert to its normal, or nonsuperconducting, state, even though the material is kept well below the transition temperature.

Suggested uses for superconducting materials include medical magnetic-imaging devices, magnetic energy-storage systems, motors, generators, transformers, computer parts, and very sensitive devices for measuring magnetic fields, voltages, or currents. The main advantages of devices made from superconductors are low power dissipation, high-speed operation, and high sensitivity.

Thermal Properties of Superconductors

Superconductivity is a startling departure from the properties of normal (i.e., nonsuperconducting) conductors of electricity. In materials that are electric conductors, some of the electrons are not bound to individual atoms but are free to move through the material; their motion constitutes an electric current. In normal conductors these so-called conduction electrons are scattered by impurities, dislocations, grain boundaries, and lattice vibrations (phonons). In a superconductor, however, there is an ordering among the conduction electrons that prevents this scattering. Consequently, electric current can flow with no resistance at all. The ordering of the electrons, called Cooper pairing, involves the momenta of the electrons rather than their positions. The energy per electron that is associated with this ordering is extremely small, typically about one thousandth of the amount by which the energy per electron changes when a chemical reaction takes place. One reason that superconductivity remained unexplained for so long is the smallness of the energy changes that accompany the transition between normal and superconducting states. In fact, many incorrect theories of superconductivity were advanced before the BCS theory was proposed.

Hundreds of materials are known to become superconducting at low temperatures. Twenty-seven of the chemical elements, all of them metals, are superconductors in their usual crystallographic forms at low temperatures and low (atmospheric) pressure. Among these are commonly known metals such as aluminum, tin, lead, and mercury and less common ones such as rhenium, lanthanum, and protactinium. In addition, 11 chemical elements that are metals, semimetals, or semiconductors are superconductors at low temperatures and high pressures. Among these are uranium, cerium, silicon, and selenium. Bismuth and five other elements, though not superconducting in their usual crystallographic form, can be made superconducting by preparing them in a highly disordered form, which is stable at extremely low temperatures. Superconductivity is not exhibited by any of the magnetic elements chromium, manganese, iron, cobalt, or nickel.

Most of the known superconductors are alloys or compounds. It is possible for a compound to be superconducting even if the chemical elements constituting it are not; examples are disilver fluoride (Ag_2F) and a compound of carbon and potassium (C_8K). Some semiconducting compounds, such as tin telluride (SnTe), become superconducting if they are properly doped with impurities.

Since 1986 some compounds containing copper and oxygen (called cuprates) have been found to have extraordinarily high transition temperatures, denoted T_c. This is the temperature below which a substance is superconducting. The properties of these high-T_c compounds are different in some respects from those of the types of superconductors known prior to 1986, which will be referred to as classic superconductors. The properties possessed by both kinds of superconductors, with attention paid to specific differences for the high-T_c materials. A further classification problem is presented by the superconducting compounds of carbon (sometimes doped with other atoms) in which the carbon atoms are on the surface of a cluster with a spherical or spheroidal crystallographic structure. These compounds, discovered in the 1980s, are called fullerenes (if only carbon is present) or fullerides (if doped). They have superconducting transition temperatures higher than those of the classic superconductors. It is not yet known whether these compounds are fundamentally similar to the cuprate high-temperature superconductors.

Transition Temperatures

The vast majority of the known superconductors have transition temperatures that lie between 1 K and 10 K. Of the chemical elements, tungsten has the lowest transition temperature, 0.015 K, and niobium the highest, 9.2 K. The transition temperature is usually very sensitive to the presence of magnetic impurities. A few parts per million of manganese in zinc, for example, lowers the transition temperature considerably.

Specific Heat and Thermal Conductivity

The thermal properties of a superconductor can be compared with those of the same material at the same temperature in the normal state. (The material can be forced into the normal state at low temperature by a large enough magnetic field).

When a small amount of heat is put into a system, some of the energy is used to increase the lattice vibrations (an amount that is the same for a system in the normal and in the superconducting state), and the remainder is used to increase the energy of the conduction electrons. The electronic specific heat (C_e) of the electrons is defined as the ratio of that portion of the heat used by the electrons to the rise in temperature of the system. The specific heat of the electrons in a superconductor varies with the absolute temperature (T) in the normal and in the superconducting state. The electronic specific heat in the superconducting state (designated C_{es}) is smaller than in the normal state (designated C_{en}) at low enough temperatures, but C_{es} becomes larger than Cen as the transition temperature T_c is approached, at which point it drops abruptly to Cen for the classic superconductors, although the curve has a cusp shape near T_c for the high-T_c superconductors. Precise measurements have indicated that, at temperatures considerably below the transition temperature, the logarithm of the electronic specific heat is inversely proportional to the temperature. This temperature dependence, together with the principles of statistical mechanics, strongly suggests that there is a gap in the distribution of energy levels available to the electrons in a superconductor, so that a minimum energy is required for the excitation of each electron from a state below the gap to a state above the gap. Some of the high-T_c superconductors provide an additional contribution to the specific heat, which is proportional to the temperature. This behaviour indicates that there are electronic states lying at low energy; additional evidence of such states is obtained from optical properties and tunneling measurements.

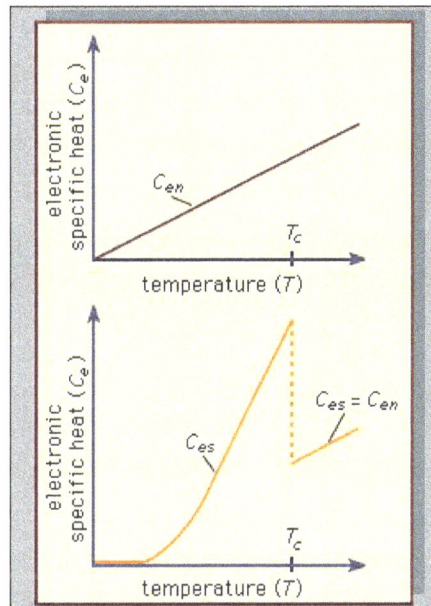

Specific heat in the normal (C_{en}) and superconducting (C_{es}) states of a classic superconductor as a function of absolute temperature. The two functions are identical at the transition temperature (T_c) and above T_c.

The heat flow per unit area of a sample equals the product of the thermal conductivity (K) and the temperature gradient ΔT: $J_Q = -K \Delta T$, the minus sign indicating that heat always flows from a warmer to a colder region of a substance.

The thermal conductivity in the normal state (K_n) approaches the thermal conductivity in the superconducting state (K_s) as the temperature (T) approaches the transition temperature (T_c) for all materials, whether they are pure or impure. This suggests that the energy gap (Δ) for each electron approaches zero as the temperature (T) approaches the transition temperature (T_c). This would also account for the fact that the electronic specific heat in the superconducting state (C_{es}) is higher than in the normal state (C_{en}) near the transition temperature: as the temperature is raised toward the transition temperature (T_c), the energy gap in the superconducting state decreases, the number of thermally excited electrons increases, and this requires the absorption of heat.

Energy Gaps

As stated above, the thermal properties of superconductors indicate that there is a gap in the distribution of energy levels available to the electrons, and so a finite amount of energy, designated as delta (Δ), must be supplied to an electron to excite it. This energy is maximum (designated Δ_0) at absolute zero and changes little with increase of temperature until the transition temperature is approached, where Δ decreases to zero, its value in the normal state. The BCS theory predicts an energy gap with just this type of temperature dependence.

According to the BCS theory, there is a type of electron pairing (electrons of opposite spin acting in unison) in the superconductor that is important in interpreting many superconducting phenomena. The electron pairs, called Cooper pairs, are broken up as the superconductor is heated. Each time a pair is broken, an amount of energy that is at least as much as the energy gap (Δ) must be supplied to each of the two electrons in the pair, so an energy at least twice as great (2Δ) must be supplied to the superconductor. The value of twice the energy gap at 0 K (which is $2\Delta_0$) might be assumed to be higher when the transition temperature of the superconductor is higher. In fact, the BCS theory predicts a relation of this type—namely, that the energy supplied to the superconductor at absolute zero would be $2\Delta_0 = 3.53$ kT_c, where k is Boltzmann's constant (1.38×10^{-23} joule per kelvin). In the high-Tc cuprate compounds, values of $2\Delta_0$ range from approximately three to eight multiplied by kT_c.

The energy gap (Δ) can be measured most precisely in a tunneling experiment (a process in quantum mechanics that allows an electron to escape from a metal without acquiring the energy required along the way according to the laws of classical physics). In this experiment, a thin insulating junction is prepared between a superconductor and another metal, assumed here to be in the normal state. In this situation, electrons can quantum mechanically tunnel from the normal metal to the superconductor if they have sufficient energy. This energy can be supplied by applying a negative voltage (V) to the normal metal, with respect to the voltage of the superconductor.

Tunneling will occur if eV—the product of the electron charge, e (-1.60×10^{-19} coulomb), and the voltage—is at least as large as the energy gap Δ. The current flowing between the two sides of the junction is small up to a voltage equal to V = Δ/e, but then it rises sharply. This provides an experimental determination of the energy gap (Δ). In describing this experiment it is assumed here that the tunneling electrons must get their energy from the applied voltage rather than from thermal excitation.

Magnetic and Electromagnetic Properties of Superconductors

Critical Field

One of the ways in which a superconductor can be forced into the normal state is by applying a magnetic field. The weakest magnetic field that will cause this transition is called the critical field (H_c) if the sample is in the form of a long, thin cylinder or ellipsoid and the field is oriented parallel to the long axis of the sample. (In other configurations the sample goes from the superconducting state into an intermediate state, in which some regions are normal and others are superconducting, and finally into the normal state.) The critical field increases with decreasing temperature. For the superconducting elements, its values (H_0) at absolute zero range from 1.1 oersted for tungsten to 830 oersteds for tantalum.

These remarks about the critical field apply to ordinary (so-called type I) superconductors. In the following section the behaviour of other (type II) superconductors is examined.

The Meissner Effect

A type I superconductor in the form of a long, thin cylinder or ellipsoid remains superconducting at a fixed temperature as an axially oriented magnetic field is applied, provided the applied field does not exceed a critical value (H_c). Under these conditions, superconductors exclude the magnetic field from their interior, as could be predicted from the laws of electromagnetism and the fact that the superconductor has no electric resistance. A more astonishing effect occurs if the magnetic field is applied in the same way to the same type of sample at a temperature above the transition temperature and is then held at a fixed value while the sample is cooled. It is found that the sample expels the magnetic flux as it becomes superconducting. This is called the Meissner effect. Complete expulsion of the magnetic flux (a complete Meissner effect) occurs in this way for certain superconductors, called type I superconductors, but only for samples that have the described geometry. For samples of other shapes, including hollow structures, some of the magnetic flux can be trapped, producing an incomplete or partial Meissner effect.

Type II superconductors have a different magnetic behaviour. Examples of materials of this type are niobium and vanadium (the only type II superconductors among the chemical elements) and some alloys and compounds, including the high-T_c compounds. As a sample of this type, in the form of a long, thin cylinder or ellipsoid, is exposed to a decreasing magnetic field that is axially oriented with the sample, the increase of magnetization, instead of occurring suddenly at the critical field sets in gradually. Beginning at the upper critical field (H_{c2}), it is completed at a lower critical field. If the sample is of some other shape, is hollow, or is inhomogeneous or strained, some magnetic flux remains trapped, and some magnetization of the sample remains after the applied field is completely removed. Known values of the upper critical field extend up to 6×10^5 oersteds, the value for the compound of lead, molybdenum, and sulfur with formula $PbMo_6S_8$.

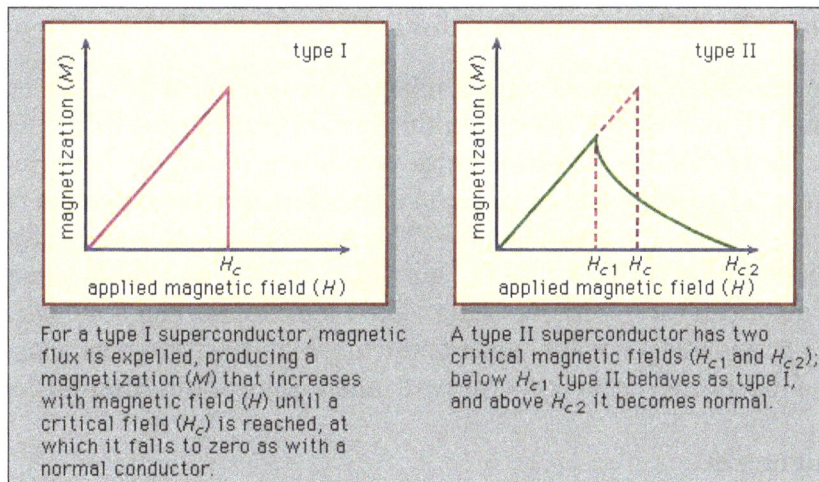

For a type I superconductor, magnetic flux is expelled, producing a magnetization (M) that increases with magnetic field (H) until a critical field (H_c) is reached, at which it falls to zero as with a normal conductor.

A type II superconductor has two critical magnetic fields (H_{c1} and H_{c2}); below H_{c1} type II behaves as type I, and above H_{c2} it becomes normal.

Magnetization as a function of magnetic field for a type I superconductor and a type II superconductor.

The expulsion of magnetic flux by type I superconductors in fields below the critical field (H_c) or by type II superconductors in fields below H_{c1} is never quite as complete as has been stated in this simplified presentation, because the field always penetrates into a sample for a small distance, known as the electromagnetic penetration depth. Values of the penetration depth for the superconducting elements at low temperature lie in the range from about 390 to 1,300 angstroms. As the temperature approaches the critical temperature, the penetration depth becomes extremely large.

High-frequency Electromagnetic Properties

The foregoing descriptions have pertained to the behavior of superconductors in the absence of electromagnetic fields or in the presence of steady or slowly varying fields; the properties of superconductors in the presence of high-frequency electromagnetic fields, however, have also been studied.

The energy gap in a superconductor has a direct effect on the absorption of electromagnetic radiation. At low temperatures, at which a negligible fraction of the electrons are thermally excited to states above the gap, the superconductor can absorb energy only in a quantized amount that is at least twice the gap energy (at absolute zero, $2\Delta_0$). In the absorption process, a photon (a quantum of electromagnetic energy) is absorbed, and a Cooper pair is broken; both electrons in the pair become excited. The photon's energy (E) is related to its frequency (v) by the Planck relation, $E = hv$, in which h is Planck's constant (6.63×10^{-34} joule second). Hence the superconductor can absorb electromagnetic energy only for frequencies at least as large as $2\Delta_0/h$.

Magnetic-flux Quantization

The laws of quantum mechanics dictate that electrons have wave properties and that the properties of an electron can be summed up in what is called a wave function. If several wave functions are in phase (i.e., act in unison), they are said to be coherent. The theory of superconductivity indicates that there is a single, coherent, quantum mechanical wave function that determines the behaviour of all the superconducting electrons. As a consequence, a direct relationship can be shown to exist between the velocity of these electrons and the magnetic flux (Φ) enclosed within any closed path inside the superconductor. Indeed, inasmuch as the magnetic flux arises because of the motion of the electrons, the magnetic flux can be shown to be quantized; i.e., the intensity of this trapped flux can change only by units of Planck's constant divided by twice the electron charge.

When a magnetic field enters a type II superconductor (in an applied field between the lower and upper critical fields, H_{c1} and H_{c2}), it does so in the form of quantized fluxoids, each carrying one quantum of flux. These fluxoids tend to arrange themselves in regular patterns that have been detected by electron microscopy and by neutron diffraction. If a large enough current is passed through the superconductor, the fluxoids move. This motion leads to energy dissipation that can heat the superconductor and drive it into the normal state. The maximum current per unit area that a superconductor can carry without being forced into the normal state is called the critical current density (J_c). In making wire for superconducting high-field magnets, manufacturers try to fix the positions of the fluxoids by making the wire inhomogeneous in composition.

Josephson Currents

If two superconductors are separated by an insulating film that forms a low-resistance junction between them, it is found that Cooper pairs can tunnel from one side of the junction to the other. Thus, a flow of electrons, called the Josephson current, is generated and is intimately related to the phases of the coherent quantum mechanical wave function for all the superconducting electrons on the two sides of the junction. It was predicted that several novel phenomena should be observable, and experiments have demonstrated them. These are collectively called the Josephson effect or effects.

The first of these phenomena is the passage of current through the junction in the absence of a voltage across the junction. The maximum current that can flow at zero voltage depends on the magnetic flux (Φ) passing through the junction as a result of the magnetic field generated by currents in the junction and elsewhere. The dependence of the maximum zero-voltage current on the magnetic field applied to a junction between two superconductors is shown in figure.

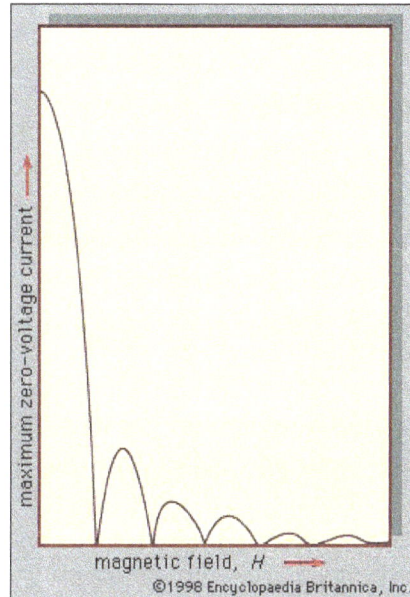

Maximum zero-voltage (Josephson) current passing through a junction
by Cooper-pair tunneling as a function of magnetic field.

A second type of Josephson effect is an oscillating current resulting from a relation between the voltage across the junction and the frequency (v) of the currents associated with Cooper pairs passing through the junction. The frequency (v) of this Josephson current is given by $v = 2eV/h$, where e is the charge of the electron. Thus, the frequency increases by 4.84×10^{14} hertz (cycles per second) for each additional volt applied to the junction. This effect can be demonstrated in various ways. The voltage can be established with a source of direct-current (DC) power, for instance, and the oscillating current can be detected by the electromagnetic radiation of frequency (v) that it generates. Another method is to expose the junction to radiation of another frequency (v') generated externally. It is found that a graph of the DC current versus voltage has current steps at values of the voltage corresponding to Josephson frequencies that are integral multiples (n) of the external frequency ($v = nv'$); that is, $V = nhv'/2e$. The observation of current steps of this type has made it possible to measure h/e with far greater precision than by any other method and has therefore contributed to knowledge of the fundamental constants of nature.

The Josephson effect has been used in the invention of novel devices for extremely high-sensitivity measurements of currents, voltages, and magnetic fields.

Low-temperature Superconductivity

In a normal metal, the electrons act relatively independently of both each other and the lattice of ions. Under certain conditions, however, the delocalized cloud can cause indirect bonding between electrons, thereby giving current enough coherence to resist any collisions or electrostatic interactions that individual electrons may experience. In this way, the only influence on the electrons is electron potential, a gradient of which is provided by an electromotive force. The result of this is a conductor without any resistance at all - a superconductor.

The temperature of the material is absolutely critical for electron bonding. The ion lattice must have such low energy that most of the valence electrons remain with their associated atoms, and the vibration of the lattice is only very slight. As a result, it is impossible to produce this state in temperatures above about 25K. The boiling point of liquid helium is around 4K, so this is ideal for use with low-temperature superconductors.

Under these conditions, when an electron is liberated from an atom's orbital, the energy of the lattice is low enough that the moving electron exerts a significant attractive force on the surrounding ion lattice, leaving a region of low electron potential in its wake. This region then attracts other electrons causing them to follow the path of the first. As a result, an indirect attraction between the two electrons is formed, pairing them over a fixed distance - about a thousand times the spacing between adjacent ions in the lattice. This attraction is only possible if both electrons have opposed spins, which reduces the electrostatic repulsion between them. In fact, the two electrons in an orbital always have opposed spins and very similar energies, and the valence orbital is destabilised under these extremely low temperatures. This means that most often the pairs will consist of electrons originating from the same orbital within a very short time of each other. This interaction between electrons is called Cooper Pairing, named for Leon Cooper who received the 1972 Nobel Prize for Physics along with John Bardeen and John Schreiffer.

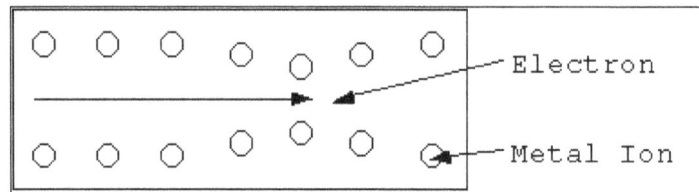

Effect of Electrons on Metallic Ion Lattice.

It is preferable for the electrons to remain in this state, separated by a distance much larger than would be the case if their energy levels dropped and they re-entered the orbital they had just left. The result is that virtually all of the delocalised cloud consists of paired electrons with energy only slightly higher than they would have in the ground state.

Not only constantly interacting with each other, but also being in the same quantum state gives the whole cloud of electrons a high degree of coherence despite every electron not being paired to every possible other. This means that even if one pair of electrons independently receive enough energy to break the Cooper pair, they will not dissociate with each other unless they also reach the next quantum energy level, bringing a far greater degree of stability to the superconductive state than might otherwise be expected.

However, the paired state is still relatively fragile, because the amount of energy required to raise the quantum state is fairly trivial. As a result, there are several considerations which ensure that a superconductor remains in this state:

- If the temperature exceeds a critical threshold, the electrons throughout the whole material will lose coherence due to increased kinetic energy. This will not only physically break the bonding within the Cooper pairs but also provide the ion lattice with enough energy to resist attraction to passing electrons, preventing broken pairs from reforming.

- The magnetic field applied must be below the critical field strength, or electrons caught in it may gain kinetic energy as a result of the motor effect. Because of this, superconductors are seldom used with alternating currents.

- Excessive charge density can lead to disruption of the lattice, causing it to be unable to mediate the interactions between pairs of electrons. This is the result of so many electrons being present that any difference in electron potential created is evened out more quickly than Cooper pairs can form. In addition, a very high current will generate a magnetic field strong enough to destroy the superconductivity.

As soon as the pairs are separated, the material reverts to being a conventional conductor once again.

High-temperature Superconductivity

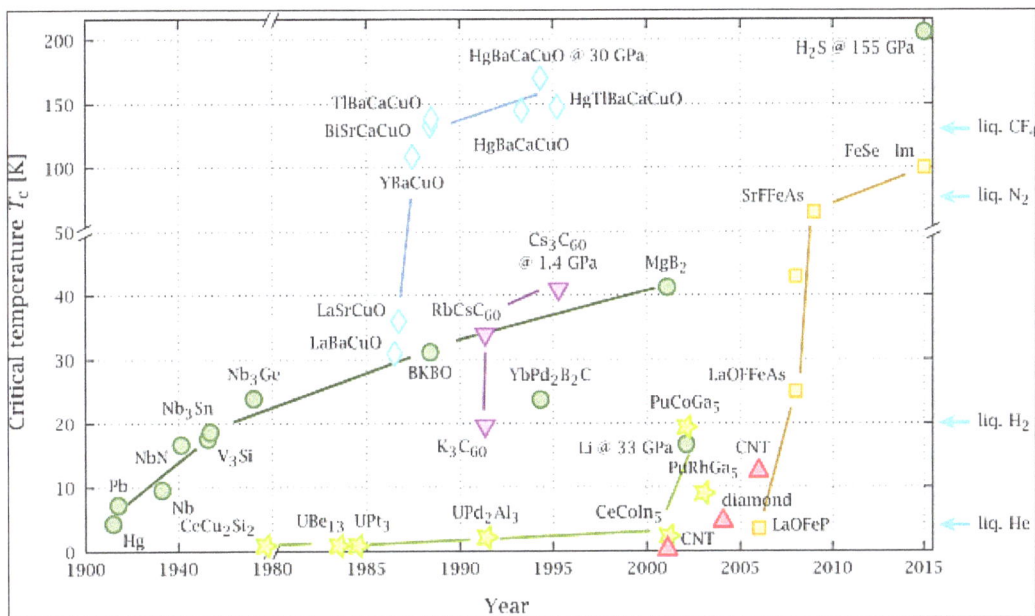

Timeline of Superconductivity - Materials and Critical Temperatures.

Discovery and Composition of High-temperature Superconductors

Ever since Kamerlingh Onnes discovered that mercury becomes superconducting at temperatures less than 4 K, scientists have been searching for superconducting materials with higher transition temperatures. Until 1986 a compound of niobium and germanium (Nb_3Ge) had the highest known transition temperature, 23 K, less than a 20-degree increase in 75 years. Most researchers expected that the next increase in transition temperature would be found in a similar metallic alloy and that the rise would be only one or two degrees. In 1986, however, the Swiss physicist Karl Alex Müller and his West German associate, Johannes Georg Bednorz, discovered, after a three-year search among metal oxides, a material that had an unprecedentedly high transition temperature of about 30 K. They were awarded the Nobel Prize for Physics in 1987, and their discovery immediately stimulated groups of investigators in China, Japan, and the United States to produce superconducting oxides with even higher transition temperatures.

These high-temperature superconductors are ceramics. They contain lanthanum, yttrium, or another of the rare-earth elements or bismuth or thallium; usually barium or strontium (both alkaline-earth elements); copper; and oxygen. Other atomic species can sometimes be introduced by chemical substitution while retaining the high-T_c properties.

Table: Transition Temperatures of Some High-T_c Super conductors.

Compound	$T_c(K)$
$Nd_{1.85}Ce_{0.15}CuO_4$	24
$La_{1.85}Sr_{0.15}CuO_4$	40
$Y Ba_2 Cu_3 O_7$	92
$Bi_2 Sr_2 Ca_2 Cu_3 O_{10}$	110
$Ti_2 Ba_2 Ca_2 Cu_3 O_{10}$	127
$Hg_2 Ba_2 Ca_2 Cu_3 O_8$	134

The table lists the member of each major family of high-T_c materials with the highest observed superconducting transition temperature. The value 134 K is the highest known T_c value. Within each family of high-T_c materials, only the subscripts (i.e., stoichiometry) vary from one compound to another. Samples in the families containing bismuth or thallium always exhibit a great deal of atomic disorder, with atoms in the "wrong" crystallographic sites and with impurity phases. It is possible that such disorder is required to make these compounds thermodynamically stable.

Structures and Properties

The compounds have crystal structures containing planes of Cu and O atoms, and some also have chains of Cu and O atoms. The roles played by these planes and chains have come under intense investigation. Varying the oxygen content or the heat treatment of the materials dramatically changes their transition temperatures, critical magnetic fields, and other properties. Single crystals of the high-temperature superconductors are very anisotropic—i.e., their properties associated with a direction, such as the critical fields or the critical current density, are highly dependent on the angle between that direction and the rows of atoms in the crystal.

If the number of superconducting electrons per unit volume is locally disturbed by an applied force (typically electric or magnetic), this disturbance propagates for a certain distance in the material; the distance is called the superconducting coherence length (or Ginzburg-Landau coherence length), ξ. If a material has a superconducting region and a normal region, many of the superconducting properties disappear gradually—over a distance ξ—upon traveling from the former to the latter region. In the pure (i.e., undoped) classic superconductors ξ is on the order of a few thousand angstroms, but in the high-T_c superconductors it is on the order of 1 to 10 angstroms. The small size of ξ affects the thermodynamic and electromagnetic properties of the high-T_c superconductors. For example, it is responsible for the cusp shape of the specific heat curve near T_c. It is also responsible for the ability of the high-T_c superconductors to remain superconducting in extraordinarily large fields—on the order of 1,000,000 gauss (100 teslas)—at low temperatures.

The high-T_c superconductors are type II superconductors. They exhibit zero resistance, strong diamagnetism, the Meissner effect, magnetic flux quantization, the Josephson effects, an electromagnetic penetration depth, an energy gap for the superconducting electrons, and the characteristic temperature dependencies of the specific heat and the thermal conductivity. Therefore, it is clear that the conduction electrons in these materials form the Cooper pairs used to explain superconductivity in the BCS theory. Thus, the central conclusions of the BCS theory are demonstrated. Indeed, that theory guided Bednorz and Müller in their search for high-temperature superconductors. It is not known, however, why the transition temperatures of these oxides are so high. It was generally believed that the members of a Cooper pair are bound together because of interactions between the electrons and the lattice vibrations (phonons), but it is unlikely that these interactions are strong enough to explain transition temperatures as high as 90 K. Most experts believe that interactions among the electrons generate high-temperature superconductivity. The details of this interaction are difficult to treat theoretically because the motions of the electrons are strongly correlated with each other and because magnetic phenomena play an important part in determining the microscopic properties of these materials. These strong correlations and magnetic properties may be responsible for unusual temperature dependencies of the electric resistivity ρ and Hall coefficient RH in the normal state (i.e., above T_c). It is observed that at temperatures above T_c the electric resistivity, although higher for superconductors than for typical metals in the normal state, is roughly proportional to the temperature T, unusually weak temperature dependence. Measurements of R_H show it to be significantly temperature-dependent in the normal state (sometimes proportional to 1/T) rather than being roughly independent of T, which is the case for ordinary materials.

Crystal Structures of High-temperature Ceramic Superconductors

The structure of high-T_c copper oxide or cuprate superconductors are often closely related to perovskite structure and the structure of these compounds has been described as a distorted, oxygen deficient multi-layered perovskite structure. One of the properties of the crystal structure of oxide superconductors is an alternating multi-layer of CuO_2 planes with superconductivity taking place between these layers. The more layers of CuO_2 the higher T_c. This structure causes a large anisotropy in normal conducting and superconducting properties, since electrical currents are carried by holes induced in the oxygen sites of the CuO_2 sheets. The electrical conduction is highly anisotropic, with a much higher conductivity parallel to the CuO_2 plane than in the perpendicular direction. Generally, Critical temperatures depend on the chemical compositions, cations substitutions and oxygen content. They can be classified as superstripes; i.e., particular realizations of superlattices at atomic limit made of superconducting atomic layers, wires, dots separated by spacer layers that give multiband and multigap superconductivity.

YBaCuO Superconductors

The first superconductor found with T_c > 77 K (liquid nitrogen boiling point) is yttrium barium copper oxide ($YBa_2Cu_3O_7$-x), the proportions of the 3 different metals in the $YBa_2Cu_3O_7$ superconductor are in the mole ratio of 1 to 2 to 3 for yttrium to barium to copper respectively. Thus, this particular superconductor is often referred to as the 123 superconductor.

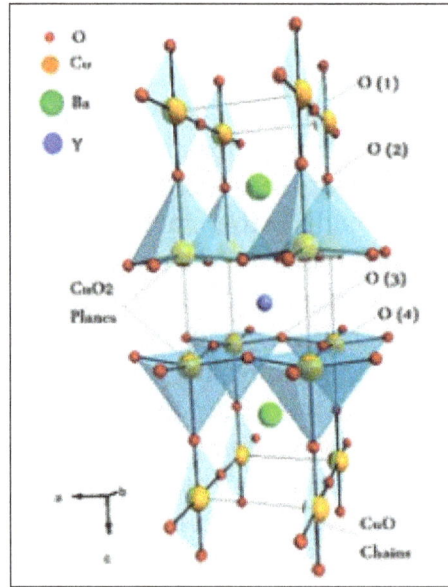

The unit cell of $YBa_2Cu_3O_7$ consists of three pseudocubic elementary perovskite unit cells. Each perovskite unit cell contains a Y or Ba atom at the center: Ba in the bottom unit cell, Y in the middle one, and Ba in the top unit cell. Thus, Y and Ba are stacked in the sequence [Ba–Y–Ba] along the c-axis. All corner sites of the unit cell are occupied by Cu, which has two different coordinations, Cu(1) and Cu(2), with respect to oxygen. There are four possible crystallographic sites for oxygen: O(1), O(2), O(3) and O(4). The coordination polyhedra of Y and Ba with respect to oxygen are different. The tripling of the perovskite unit cell leads to nine oxygen atoms, whereas $YBa_2Cu_3O_7$ has seven oxygen atoms and, therefore, is referred to as an oxygen-deficient perovskite structure. The structure has a stacking of different layers: $(CuO)(BaO)(CuO_2)(Y)(CuO_2)(BaO)(CuO)$. One of the key feature of the unit cell of $YBa_2Cu_3O_7$-x (YBCO) is the presence of two layers of CuO_2. The role of the Y plane is to serve as a spacer between two CuO_2 planes. In YBCO, the Cu–O chains are known to play an important role for superconductivity. T_c is maximal near 92 K when x $\approx$ 0.15 and the structure is orthorhombic. Superconductivity disappears at x $\approx$ 0.6, where the structural transformation of YBCO occurs from orthorhombic to tetragonal.

Bi-, Tl- and Hg-based High-T_c Superconductors

The crystal structure of Bi-, Tl- and Hg-based high-T_c superconductors are very similar. Like YBCO, the perovskite-type feature and the presence of CuO_2 layers also exist in these superconductors. However, unlike YBCO, Cu–O chains are not present in these superconductors. The YBCO superconductor has an orthorhombic structure, whereas the other high-T_c superconductors have a tetragonal structure.

Bi-based High-T_c Superconductors (Bi–Sr–Ca–Cu–O)

The Bi–Sr–Ca–Cu–O system has three superconducting phases forming a homologous series as $Bi_2Sr_2Ca_{n-1}Cu_nO_{4+2n+x}$ (n = 1, 2 and 3). These three phases are Bi-2201, Bi-2212 and Bi-2223, having transition temperatures of 20, 85 and 110 K, respectively, where the numbering system represent number of atoms for Bi, Sr, Ca and Cu respectively. The two phases have a tetragonal structure which consists of two sheared crystallographic unit cells. The unit cell of these phases has double Bi–O planes which are stacked in a way that the Bi atom of one plane sits below the oxygen atom of the next consecutive plane. The Ca atom forms a layer within the interior of the CuO_2 layers in both Bi-2212 and Bi-2223; there is no Ca layer in the Bi-2201 phase. The three phases differ with each other in the number of CuO_2 planes; Bi-2201, Bi-2212 and Bi-2223 phases have one, two and three CuO_2 planes, respectively. The c axis of these phases increases with the number of CuO_2 planes. The coordination of the Cu atom is different in the three phases. The Cu atom forms an octahedral coordination with respect to oxygen atoms in the 2201 phase, whereas in 2212, the Cu atom is surrounded by five oxygen atoms in a pyramidal arrangement. In the 2223 structure, Cu has two coordinations with respect to oxygen: one Cu atom is bonded with four oxygen atoms in square planar configuration and another Cu atom is coordinated with five oxygen atoms in a pyramidal arrangement.

Tl-based High-T_c Superconductors (Tl–Ba–Ca–Cu–O)

The first series of the Tl-based superconductor containing one Tl–O layer has the general formula $TlBa_2Ca_{n-1}CunO_{2n+3}$, whereas the second series containing two Tl–O layers has a formula of $Tl_2Ba_2Ca_{n-1}CunO_{2n+4}$ with n = 1, 2 and 3. In the structure of $Tl_2Ba_2CuO_6$ (Tl-2201), there is one CuO_2 layer with the stacking sequence (Tl–O) (Tl–O) (Ba–O) (Cu–O) (Ba–O) (Tl–O) (Tl–O). In $Tl_2Ba_2CaCu_2O_8$ (Tl-2212), there are two Cu–O layers with a Ca layer in between. Similar to the Tl2Ba2CuO6structure, Tl–O layers are present outside the Ba–O layers. In $Tl_2Ba_2Ca_2Cu_3O_{10}$ (Tl-2223), there are three CuO_2 layers enclosing Ca layers between each of these. In Tl-based superconductors, Tc is found to increase with the increase in CuO_2 layers. However, the value of T_c decreases after four CuO_2 layers in $TlBa_2Ca_{n-1}CunO_{2n+3}$, and in the $Tl_2Ba_2Ca_{n-1}CunO_{2n+4}$ compound, it decreases after three CuO_2 layers.

Hg-based High-T_c superconductors (Hg–Ba–Ca–Cu–O)

The crystal structure of $HgBa_2CuO_4$ (Hg-1201), $HgBa_2CaCu_2O_6$ (Hg-1212) and $HgBa_2Ca_2Cu_3O_8$ (Hg-1223) is similar to that of Tl-1201, Tl-1212 and Tl-1223, with Hg in place of Tl. It is noteworthy that the Tc of the Hg compound (Hg-1201) containing one CuO_2 layer is much larger as compared to the one-CuO_2-layer compound of thallium (Tl-1201). In the Hg-based superconductor, T_c is also found to increase as the CuO_2 layer increases. For Hg-1201, Hg-1212 and Hg-1223, the values of T_c are 94, 128 and the record value at ambient pressure 134 K, respectively, as shown in table below. The observation that the T_c of Hg-1223 increases to 153 K under high pressure indicates that the T_c of this compound is very sensitive to the structure of the compound.

Table: Critical temperature (T_c), crystal structure and lattice constants of some high-T_c superconductors.

Formula	Notation	T_c (K)	No. of Cu-O planes in unit cell	Crystal structure
$YBa_2Cu_3O_7$	123	92	2	Orthorhombic
$Bi_2Sr_2CuO_6$	Bi-2201	20	1	Tetragonal
$Bi_2Sr_2CaCu_2O_8$	Bi-2212	85	2	Tetragonal
$Bi_2Sr_2Ca_2Cu_3O_6$	Bi-2223	110	3	Tetragonal

$Tl_2Ba_2CuO_6$	Tl-2201	80	1	Tetragonal
$Tl_2Ba_2CaCu_2O_8$	Tl-2212	108	2	Tetragonal
$Tl_2Ba_2Ca_2Cu_3O_{10}$	Tl-2223	125	3	Tetragonal
$TlBa_2Ca_3Cu_4O_{11}$	Tl-1234	122	4	Tetragonal
$HgBa_2CuO_4$	Hg-1201	94	1	Tetragonal
$HgBa_2CaCu_2O_6$	Hg-1212	128	2	Tetragonal
$HgBa_2Ca_2Cu_3O_8$	Hg-1223	134	3	Tetragonal

Fe-based High-T_c Superconductors

Iron-based superconductors contain layers of iron and a pnictogen—such asarsenic or phosphorus—or a chalcogen. This is currently the family with the second highest critical temperature, behind the cuprates. Interest in their superconducting properties began in 2006 with the discovery of superconductivity in LaFePO at 4 K and gained much greater attention in 2008 after the analogous material LaFeAs(O,F) was found to superconduct at up to 43 K under pressure.

Since the original discoveries several families of iron-based superconductors have emerged:

- LnFeAs(O,F) or $LnFeAsO_{1-x}$ with T_c up to 56 K, referred to as 1111 materials. A fluoride variant of these materials was subsequently found with similar T_c values.

- $(Ba,K)Fe_2As_2$ and related materials with pairs of iron-arsenide layers, referred to as 122 compounds. T_c values range up to 38 K. These materials also superconduct when iron is replaced with cobalt.

- LiFeAs and NaFeAs with T_c up to around 20 K. These materials superconduct close to stoichiometric composition and are referred to as 111 compounds.

- FeSe with small off-stoichiometry or tellurium doping.

Most undoped iron-based superconductors show a tetragonal-orthorhombic structural phase transition followed at lower temperature by magnetic ordering, similar to the cuprate superconductors. However, they are poor metals rather than Mott insulators and have five bands at the Fermi surface rather than one. The phase diagram emerging as the iron-arsenide layers are doped is remarkably similar, with the superconducting phase close to or overlapping the magnetic phase. Strong evidence that the T_c value varies with the As-Fe-As bond angles has already emerged and shows that the optimal T_c value is obtained with undistorted $FeAs_4$ tetrahedra. The symmetry of the pairing wavefunction is still widely debated, but an extended s-wave scenario is currently favoured.

Superconducting Phase Transition

In superconducting materials, the characteristics of superconductivity appear when the temperature T is lowered below a critical temperature T_c. The value of this critical temperature varies from material to material. Conventional superconductors usually have critical temperatures ranging from less than 1K to around 20K. Solid mercury, for example, has a critical temperature of 4.2K. As of 2001, the highest critical temperature found for a conventional superconductor is 39K for magnesium diboride (MgB_2), although this material displays enough exotic properties that there is doubt about classifying it as a "conventional" superconductor. Cuprate superconductors can have much higher critical temperatures: $YBa_2Cu_3O_7$, one of the first cuprate superconductors to be discovered, has a critical temperature of 92K, and mercury-based cuprates have been found with critical temperatures in excess of 130K. The explanation for these high critical temperatures remains unknown. (Electron pairing due to phonon exchanges explains superconductivity in conventional superconductors, while it does not explain superconductivity in the newer superconductors that have a very high T_c).

The onset of superconductivity is accompanied by abrupt changes in various physical properties, which is the hallmark of a phase transition. For example, the electronic heat capacity is proportional to the temperature in the normal (non-superconducting) regime. At the superconducting transition, it suffers a discontinuous jump and thereafter ceases to be linear. At low temperatures, it varies instead as $e^{-\alpha/T}$ for some constant α. (This exponential behavior is one of the pieces of evidence for the existence of the energy gap).

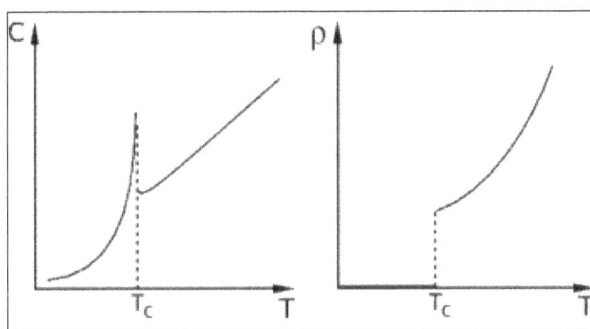

Behavior of heat capacity (C) and resistivity (ρ) at the superconducting phase transition.

The order of the superconducting phase transition is still a matter of debate. It had long been thought that the transition is second-order, meaning there is no latent heat. However, recent calculations have suggested that it may actually be weakly first-order due to the effect of long-range fluctuations in the electromagnetic field.

Thermodynamics of the Superconducting Transition

The variation of specific heat with temperature is often a good probe of phase transitions in matter. Historically, it is Ehrenfest who first classified phase transitions based on the variation of the thermodynamic free energy with some state variable such as temperature. The order of a transition was defined as the lowest derivative of free energy (with respect to some variable) that was discontinuous at the transition. If the first derivative of free energy were discontinuous (such as the case of a solid-liquid transition where the density is discontinuous), then the transition is called first order. In the case of ferromagnetic transition of Fe for example, the susceptibility (i.e., the second derivative of free energy with field) is discontinuous and one would classify this as a second order phase transition. However, there are many cases in nature where rather than discontinuous jumps in thermodynamic variables, there is a divergence such as in the heat capacity of a superconductor. Over the decades, changes in these criteria have been proposed to accommodate such cases. The modern classification of phase transitions is based on the existence or lack thereof of a latent heat. If a phase transition involves a latent heat, i.e., the substance absorbs or releases heat without a change in temperature, and then it is called a 1st order phase transition. In the absence of a latent heat, the phase transition is a 2nd order transition. Landau gave a theory of 2nd order phase transitions and its application to superconductors will be discussed later in these lectures.

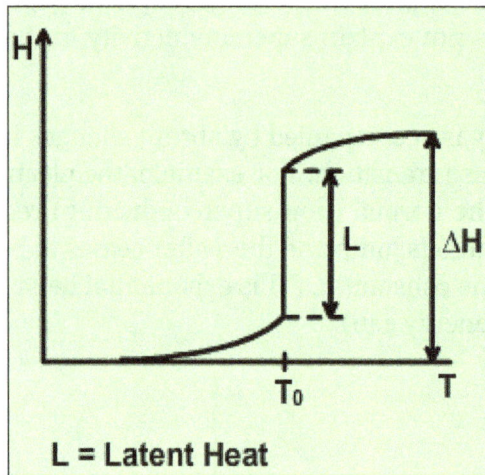

The variation of the enthalpy in the vicinity of a first order phase transition.

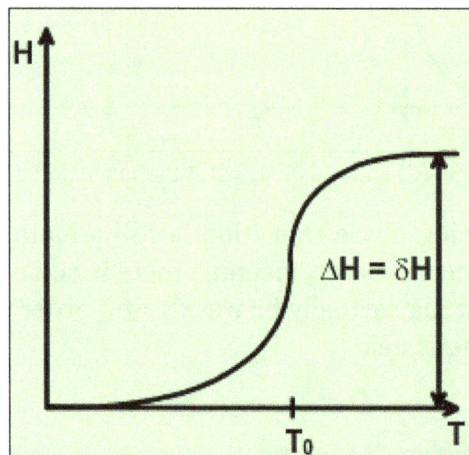

The variation of the enthalpy in the vicinity of a non-first order phase transition.

The figure shows the schematic variation of enthalpy in the case of a transition involving a latent heat. For a type I superconductor, in general, there is an entropy change at the transition temperature (and therefore a latent heat) making the transition 1st order. However, in zero magnetic field, the entropy change is zero and hence the transition is 2nd order. In the normal state, the electronic contribution to the heat capacity is linear in temperature. The heat capacity exhibits a jump at and at T_c and at lower temperatures, it falls with exponential temperature dependence. The exponential dependence is due to the opening up of a gap in the excitation spectrum. Signatures of a gap are seen in various other properties such as thermal conductivity, current-voltage characteristics, etc.

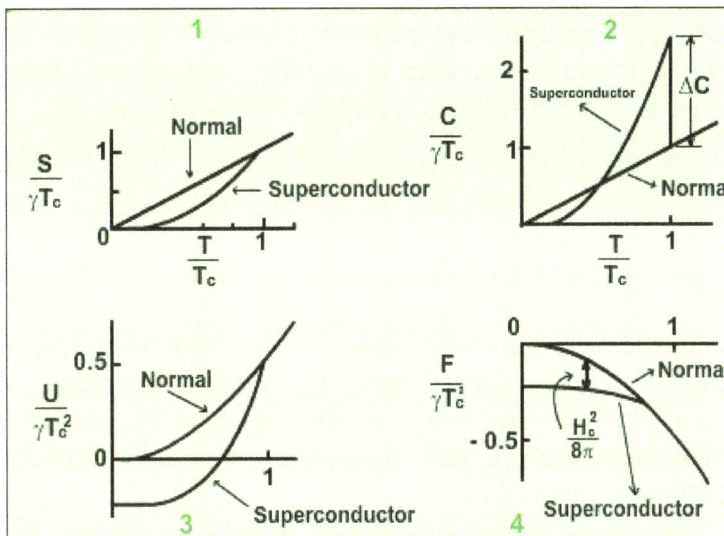

Variation of X with temperature for a normal metal and a Superconducting
- 1. Entropy; 2. Heat capacity; 3. Internal Energy; 4. Free Energy.

The accompanying figures contrast the variation with temperature of some basic thermodynamic quantities such as the entropy S, the internal energy U, the heat capacity C and the Helmholtz free energy F.

Basic Thermodynamics and Magnetism

We will first review the basic thermodynamics related to magnetic materials. Consider a solenoid having N/L turns per unit length. A long cylindrical sample (for this choice the demagnetization field negligible) is placed in the solenoid. The field in the sample is $\vec{H} = \dfrac{NI}{L}\hat{z}$ if a current I flows in the solenoid. The total work done on increasing the current from I to $I + dI$ is (finally, a positive work is done by the sources in increasing the energy of the sample and the vacuum).

$$dW = -N\varepsilon I dt$$
$$= N\frac{d\phi}{dt} I dt$$
$$= N I d\phi$$
$$= N\, A I dB$$
$$= \frac{V}{4\pi}\vec{H}\cdot d\vec{B}$$
$$= V\left(\vec{H}\cdot d\vec{M} + \vec{H}\cdot d\vec{H}\right).$$

The first term is the magnetic work done on the sample. If the sample were not present, work would still have been done to increase the electromagnetic field energy and that is the second term. Our convention will be to include ONLY the magnetic work done on the sample and not on the vacuum. The first law of thermodynamics then reads:

$$dU = TdS + V\vec{H}\cdot d\vec{M}$$

This is analogous to the gas equation where the work done on the gas is $-PdV$. The field $\overrightarrow{H}$ is similar to $-P$ and the magnetisation is similar to the volume of the gas. Therefore we can think of the internal energy as a function of the entropy and the magnetisation with,

$$T = \frac{\partial U}{\partial S}$$

$$H = \frac{1}{V}\frac{\partial U}{\partial M}.$$

However, S and M are not very convenient variables to work with from a practical viewpoint since what we control externally is the current and hence H and rather the temperature than the entropy. Therefore, it will be useful to write down an energy equation expressed in terms of changes in applied magnetic field and temperature. For a magnetic system, the Helmholtz and the Gibbs free energy can then be written down as follows:

$$F(T,M) = U - TS$$

and

$$G(T,H) = U - TS - V\overrightarrow{H}\cdot\overrightarrow{M}.$$

Therefore the appropriate free energy to consider is the Gibbs free energy (which is expressed in terms of changes in temperature and magnetic field) and the entropy and magnetisation can be calculated from its derivatives.

$$dG = dU - TdS - Sdt - V\overrightarrow{H}\cdot d\overrightarrow{M} - V\overrightarrow{M}\cdot d\overrightarrow{H}$$

$$= -SdT - V\overrightarrow{M}\cdot d\overrightarrow{H}$$

$$S = -\frac{\partial G}{\partial T},$$

$$M = -\frac{1}{V}\frac{\partial G}{\partial H}.$$

From the Gibbs free energy we can also write the Helmholtz free energy $F = G + V\overrightarrow{H}\cdot\overrightarrow{M}$ as also the internal energy U.

Superconductivity in Heavily Boron-doped Siliconcarbide

The possibility to achieve a superconducting phase in wide-band-gap semiconductors was suggested in 1964 by Cohen in Ge and GeSi. Right after the prediction, several semiconductor-based compounds were indeed found to be superconducting at rather low temperatures and high doping concentrations. In the last decade, superconductivity was found in doped silicon clathrates crystallizing in a covalent tetrahedral sp^3 network with a bond length similar to that in diamond. In 2004, type-II superconductivity was found in highly boron-doped diamond (C : B), the cubic carbon modification with a large band gap. The boron (hole) concentration of the sample was reported to be about $n = 1.8 \times 10^{21} cm^{-3}$ with a critical superconducting transition temperature $T_c \approx 4.5$ K and an upper critical field strength H_{c2} ≈ 4.2 T. At higher doping concentrations $n = 8.4 \times 10^{21} cm^{-3}$, T_c was found to increase to about 11.4 K and H_{c2} to about 8.7 T. In 2006, type-II superconductivity was discovered in its next-period neighbor in the periodic system cubic silicon (Si : B) at boron concentrations of about $n = 2.8 \times 10^{21} cm^{-3}$. However, the critical temperature is only 0.4 K and the upper critical field 0.4 T_3.

In 2007, we found superconductivity in the stoichiometric composition of carbon and silicon: heavily boron-doped silicon carbide (SiC : B). One interesting difference between these three superconducting systems is the well-known polytypism in SiC meaning that SiC exhibits various ground states of slightly different energy. More than 200 such structural modifications are reported. There is only one cubic 'C' modification labeled as 3C-SiC (zincblende = diamond structure with two different elements) or β-SiC. All other observed unit cells are either hexagonal 'H' (wurtzite or wurtzite related) or rhombohedral 'R', labeled as mH-SiC or α-SiC and mR-SiC, respectively. The variable m indicates the number of carbon and silicon bilayers which are needed to form the unit cell. The most important hexagonal modifications are 2H-SiC, which is the only pure hexagonal polytype, and 4H- and 6H-SiC, which consist of cubic and hexagonal bonds. We note here that all SiC polytypes break inversion symmetry which is known to give rise to quite unconventional superconducting scenarios, e. g. in heavy-fermion compounds, in contrast to the inversion-symmetry conserving systems C : B and Si : B. However, we do not believe that any exotic superconducting scenario applies to SiC : B because of the comparably light elements carbon and silicon.

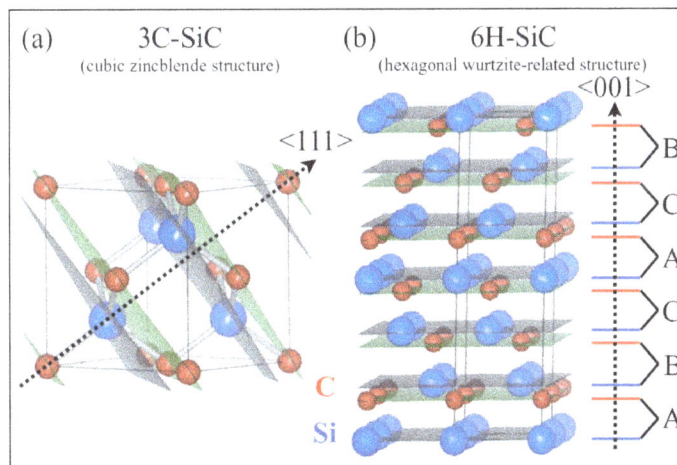

Figure show (a) Unit cell of cubic 3C-SiC. The planes mark the three C–Si bilayers forming the unit cell (stacking sequence: ABC – . . . along h111i (dotted arrow)). The tetrahedral bond alignment of diamond is emphasized demonstrating the close relation to that structure. (b) Four unit cells of hexagonal 6H-SiC. The six bilayers needed for the unit cell are again denoted by planes (stacking sequence ABCACB – . . . along h001i (dotted arrow)). For the drawings the software Vesta was used.

Superconducting Properties of SiC B

Here, we used a multiphase polycrystalline boron-doped SiC sample which contained three different phase fractions: 3C-SiC, 6H-SiC, and unreacted Si. The charge-carrier concentration of this particular sample was estimated to 1.91×10^{21} holes cm^{-3}. The critical temperature, at which we observe a sharp transition in resistivity and ac susceptibility, is ~1.45 K. The critical field strength amounts to $H_c \approx 115$ Oe, much lower than those of the two parent compounds C : B and Si : B. A big surprise was the finding that SiC : B is a type-I superconductor as indicated by a clear hysteresis between data (resistivity and ac susceptibility) measured upon cooling from above T_c to the lowest accessible temperature and a subsequent warming run in different applied external DC magnetic fields. This is in clear contrast to the reported type-II behaviour of C : B and Si : B. Another surprise is the low residual resistivity $\rho 0 = 60\mu$ cm at T_c, which is unexpected for an impure semiconductor-based system, i. e., for a multiphase polycrystalline sample. Above T_c, the system features a metallic-like temperature dependence with a positive slope of $d\rho/dT$ in the whole temperature range up to room temperature and a residual resistivity ratio RRR = $\rho(300$ K$)/\rho 0$ of about 10. These observations are again in contrast, especially to C : B, which exhibits a more or less temperature independent resistivity with $\rho 0 \approx 2500\mu$ cm and RRR ≈ 1. In a subsequent specific-heat study, using the same sample, we found a very small normal-state Sommerfeld parameter $\gamma n \approx 0.29$ mJ mol^{-1} K^{-1}. Moreover, we could clearly demonstrate that SiC : B is a bulk superconductor, as indicated by a specific-heat jump at about 1.45 K coinciding with the critical temperature T_c estimated from resistivity and ac susceptibility data. The jump in the specific heat is rather broad reflecting the multiphase polycrystalline character of the sample used. In addition, there is a third remarkable surprise. The electronic specific heat cel/T exhibits strictly linear temperature dependence below its jump down to the lowest so-far accessed temperature ~0.35 K and extrapolates almost identical to 0 for T → 0. The jump height is estimated to /γn $T_c \approx 1$, which is only 2/3 of the expectation in the Bardeen–Cooper–Schrieffer theory (BCS) of a weak-coupling superconductor and is close to the value theoretically expected for a superconducting gap with nodes. However, strictly linear temperature dependence is only expected well below T_c, where the superconducting gap is nearly temperature independent. When approaching T_c, the specific heat should deviate from cel/T $\propto$ T due to the reduction of the gap magnitude. We note here that the assumption of a BCS-like scenario with a residual contribution to the specific heat γres, e.g., due to non-superconducting parts of the sample, yields a reasonable description of the data, too, with γres ≈ 0.14 mJ mol^{-1} K^{-1}. The jump height in this scenario is 1.48, almost matching the BCS expectation of 1.43. In the description of the specific heat, assuming a linear cel/T, no residual contribution is needed. A respective fit to the data yields γres ≈ 0 mJ mol^{-1} K^{-1}, as suggested by the almost perfect extrapolation of the data down to 0 for T → 0. These results for γres for the two approaches imply a superconducting volume fraction of about 100% for the nodal gap scenario and about 50% for the BCS-like scenario.

The hexagonal phase fraction is superconducting and exhibits a similar linear temperature dependence of the electronic specific heat cel/T in the superconducting state and also a reduced jump height.

Superconductivity in Iron Pnictides

Structure and Electronic Structure

The structural motif of the iron-pnictide and chalcogenide superconductors is the occurrence of Fe square planes with Fe in a nominally divalent state, and having a tetrahedral coordination by P, As, Se, Te or alloys including alloys with S. This is illustrated along with the unit cell in figure. Aside from this common feature, a very wide variety of Fe-based superconductors have been discovered since the initial report of high T_c superconductivity in 2008. Here we focus on the features that are common to this family.

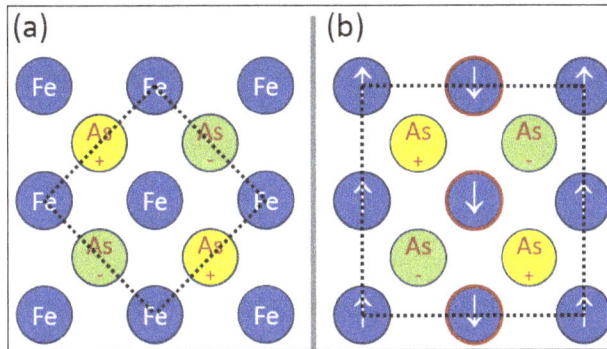

Figure show (a) Structure of the FeAs layers in iron-pnictide superconductors. The Fe atoms are in a plane with As (or chalcogens) above (denoted +) or below (denoted −) to form a tetrahedral coordination. The dotted lines show the unit cell. Note that the cell contains two iron atoms. (b) Spin density wave magnetic structure. Note the doubling of the cell and the symmetry lowering from tetragonal to orthorhombic.

Density functional calculations confirmed that Fe is divalent in these compounds and showed an electronic structure with a rather high density of states (DOS) at the Fermi level, derived mainly from Fe d states, $N(EF) \approx 2$ eV^{-1} per Fe both spins depending on the material. This in itself places the compounds near itinerant magnetism. There is hybridization between the Fe d states and the p states of the coordinating ligands (As, P, S, Se, Te), but invariably the main ligand bands are located below the Fe d bands generally at binding energies higher than 2 eV. This is in accord with spectroscopic experiments.

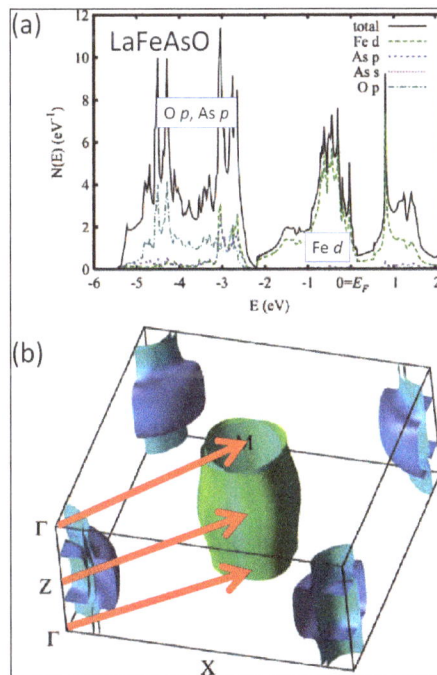

(a) Density of states and projections of LaFeAsO as obtained in density functional calculations. (b) Fermi surface of LaFeAsO. The heavy arrows indicate the nesting of the electron and hole sheets. Γ is the zone center, (0, 0, 0), X is (1/2, 0, 0), M is (1/2, 1/2, 0) and Z is (0, 0, 1/2).

Turning to the shape of the Fe d derived DOS, EF in these compounds invariably lies on the lower energy side of a pronounced dip in the DOS. This dip is at an electron count of six d electrons per Fe. In contrast, a tetrahedral crystalfield scheme would split the Fe d states into a lower lying eg manifold containing four electrons and a higher t_{2g} manifold with six electrons. This difference between the actual shape of the DOS and the crystal field scheme reflects the importance of direct Fe-Fe hopping in forming the band structure, and is a consequence of the crystal structure, specifically the fact that the structure is built from edge sharing coordination polyhedra rather than the corner sharing octahedra characteristic of the cuprates. The Fe-Fe distance in the Fe-based superconductors is $\approx$ 2.8-2.9 Å, in contrast to the nearly 4 Å in cuprates.

The DOS shows another key dierence from cuprates. In cuprates, a single $d_{x^2} - y^2$ orbital plays the dominant role in the electronic structure near EF, while in the Fe-based superconductors, the d-shell is open and all the d orbitals are involved. This is important from the point of view of the Mott physics. The multi-orbital nature of the low energy electronic structure opens channels for inter-orbital charge fluctuations that work against the Mott state. This was quantified by Gunnarsson and co-workers, who derived an approximate factor $1/\sqrt{N}$, to be multiplied by U/W when assessing the proximity of a material to a Mott transition (here U is the effective Coulomb repulsion, i.e. the Hubbard U, W is the band width and N is the number of orbitals).

Within a strongly correlated picture, the effect of the Hubbard U is generally to shift d spectral weight away from the EF to the Hubbard bands. This is the case in cuprates, but not in Fe-based superconductors, as was shown early on by X-ray absorption and photoemission experiments. There is a large Fe d spectral weight near the Fermi level, a renormalization of the d bands (by a factor of $\approx$ 2) and no Hubbard bands. The presence of electron-electron correlation, but not the Mott physics normally associated with a large Hubbard U. What is seen is a rearrangement of the spectral weight within the d bands, which can come both from the Hund rule coupling J and the Hubbard parameter U, but no Hubbard bands. Importantly, in spite of the chemical diversity of this family of materials, no Mott insulating state has been found, strongly implying that the Fe-based superconductors are not in proximity to a Mott state. Instead the electronic structures appear to be more characteristic of intermetallic compounds than correlated oxides. One reflection of this is the fact that while alloying on the Cu site in cuprates is highly destructive to superconductivity, in the Fe-based materials high temperature superconductivity can be induced by alloying on the Fe site by other metals, such as Co in $BaFe_{2-x}CoxAs_2$ and $SrFe_{2-x}CoxAs_2$, and even with replacement of as much as 40% of the Fe by Ru in $SrFe_{2-x}RuxAs_2$, x = 0.8.

Fermi Surface and Nesting

Superconductivity is fundamentally instability of the Fermi surface. Interestingly, while the band structure is metallic and the density of states is high, the electronic structure at E_F consists of relatively small disconnected Fermi surfaces, as illustrated in figure. These are hole cylinders around the zone center, and electron cylinders around the zone corner.

The details vary between the different materials and as a function of doping. However, invariably there are two electron-like sections at the zone corner. These have the shape of two intersecting elliptical sections that are derived from d_{xz}, d_{yz} states on the inner part and dxy orbitals on the outer lobes. Also invariably, there are hole cylinders derived from d_{xz}, d_{yz} orbitals at the zone center. There is also generally k_z-dependent participation of other orbitals, specifically the and dxy orbitals, at the zone center, either through hybridization or extra Fermi surface sheets. The hole sheets are generally heavier (lower Fermi velocity, less contribution to the conductivity) and more three-dimensional than the electron sheets. These basic features have been confirmed experimentally using angle-resolved photoemission (ARPES) by several groups working on the various materials.

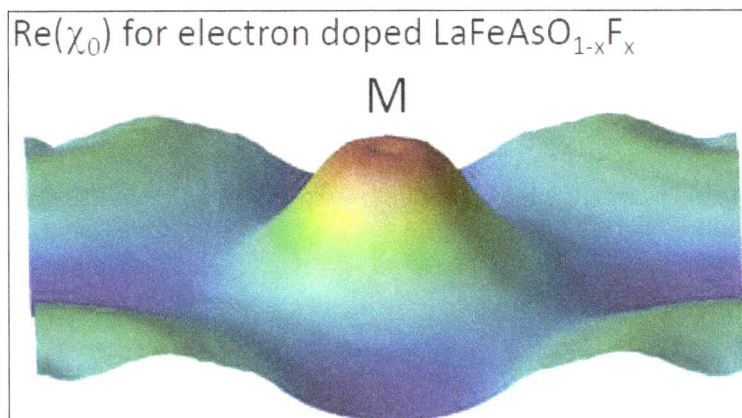

Real part of the bare (non-interacting) susceptibility of electron doped LaFeAsO.

Since both the main electron and hole sheets share similar orbital character, and have similar shapes (cylinders), one can anticipate nesting. This in fact is the case as was shown by calculations of the Lindhard function. Figure shows the non-enhanced real part of the Lindhard function for electron doped LaFeAsO. As may be seen, there is a broad peak of width comparable to the Fermi surface sizes, centered at the zone corner. With enhancement, this peak is high enough to lead to magnetic ordering, which would then be of spin-density-wave (SDW) type.

Magnetic order is observed experimentally for undoped LaFeAsO, and also for many but not all of the other compounds when undoped, one exception being FeSe. This is already different from the cuprates. The cuprates show antiferromagnetism and insulating behavior in all the undoped compounds, and superconductivity does not exist without doping. In contrast, the Fe-based superconductors show a competition between antiferromagnetic ordering and superconductivity, but there are a variety of ways of suppressing the magnetic order - doping as in the cuprates, but also isovalent alloying on the Fe or ligand sites, pressure etc. - and all of these lead to superconductivity. The particular magnetic order that is observed in proximity to the superconducting phases is that shown in figure, i.e. that corresponding to zone corner instability, with a doubling of the unit cell as shown.

Another related difference from the cuprates is the intimate connection between the antiferromagnetic and superconducting phases. Ning and co-workers measured the NMR relaxation rate as a function of temperature for a series of samples with different doping levels. They see excess relaxation due to spin fluctuations that grows as T is reduced, and which evolves continuously going from undoped, antiferromagnetic samples, to fully doped superconducting samples. Finally, the nature of the antiferromagnetic phases of the cuprates and Fe-based superconductors is very different. The antiferromagnetic phases of the cuprates are the Mott insulators. In contrast, the antiferromagnetic phases of the Fe-based superconductors are unambiguously metals, as seen both by bands dispersing through the Fermi energy in ARPES, and the observation of quantum oscillations in the magnetic phase.

There has been some debate about the origin of the magnetic phase in the Fe-based superconductors, specifically whether the instability is driven by the Fermi surface nesting or by short range superexchange interactions. This point has been discussed in detail by Johannes and Mazin, who argued that the moment formation in the Fe-based superconductors is largely driven by on-site Hund's coupling and that the inter-site interactions are mediated both by electronic states close to EF, as in an SDW, and by deeper states. This was based on density functional calculations for various materials using the experimental crystal structures. One problem with this is that such calculations strongly overestimate the magnetic moments of these materials, probably because of beyond mean field renormalizations by spin fluctuations. In any case, this characterization does capture the main features, specifically (1) the SDW type order strongly reconstructs the Fermi surface, (2) there are also rearrangements of bands away from EF, and (3) other orders are seen in this family of materials such as in $TlFe_2Se_2$ and FeTe.

Spin Fluctuations and Superconductivity

Fe-based superconductivity cannot be explained by a standard electron-phonon mechanism. Therefore, other mechanisms need to be considered. Since the materials are in proximity to magnetism, it is natural to ask whether spinfluctuation induced pairing is responsible. As shown by Berk and Schrieffer in the context of Pd, spin fluctuations are a repulsive interaction in a singlet channel. This means that they can only stabilize a superconducting state that has sign changes in the order parameter over the Fermi surface.

The pairing interaction due to spin fluctuations is closely related to the real part of the susceptibility χ and is negative. This pairing interaction will be strong and negative at the antiferromagnetic wave vector, which is the vector connecting the electron and hole Fermi surfaces. This then favors an order parameter that changes sign between these sheets. This interaction will be highly unfavorable for a pairing channel where the order parameter on the electron and hole sheets has the same sign, much in the same way that nearness to ferromagnetism is unfavorable for standard s-wave superconductivity in Pd. This observation led to the prediction of a sign changing s-wave order parameter in the Fe-based superconductors by Mazin and co-workers, and subsequently by Kuroki and co-workers. This sign changing s-wave state, denoted s±, has average order parameters of opposite sign on the electron and hole sheets of the Fermi surface but has the same symmetry as a standard s-wave state.

It should be noted that while this state has on average opposite order parameters on the two sheets and could be fully gapped, it is not necessarily nodeless. In fact, even simple Coulomb repulsion can favor a state where there are accidental nodes on the Fermi surface. From an experimental perspective the simplest way to distinguish the s± from a standard s-wave state or from other states such as d-wave is through coherence factors. One is the NMR Hebel-Slichter peak which is suppressed with an s± order parameter but not with a standard s-wave, and another is a neutron spin resonance, which should occur for this state at the 2D nesting vector, (π, π).

This resonance has been observed at the nesting wave vector showing that there is a sign change between the Fermi surfaces sections separated by it, i.e. as expected in the s± case. Interestingly, in doped $BaFe_2As_2$, which has a noticeable corrugation of the Fermi surface along the k_z direction, the resonance also shows k_z dependence depending on doping and presumably reflecting k_z dispersion of the spin-fluctuations.

Within a mean field picture, SDW magnetic ordering will occur when the bare (non-enhanced) susceptibility, $Re(\chi_o)$ exceeds a threshold value at the ordering wave vector, q, so that the RPA enhanced susceptibility $\chi(q) = \chi_0(q) / \left[1 - I(q) \chi_0(q) \right]$, diverges. What matters therefore is the magnitude at a specific q. Beyond the RPA level, spin fluctuations work against ordering. The extent of this suppression is related to an integral of the imaginary part of the susceptibility over wave vector and energy by the fluctuation dissipation theorem. Qualitatively, this reflects the intuitive result that competition between different magnetic states works against ordering. As such, for a given peak value of χ_o, a sharp peak is more favorable for magnetic ordering than a broad peak.

In contrast, the BCS gap equation involves an integral of the order parameter with the pairing interaction (related to $Re(\chi)$), i.e. for the s± state, the integral of over the region of q that can connect the electron and hole Fermi surfaces (a region set by the Fermi surface size). Thus, for superconductivity a broad peak with a large weight is much better than a narrow peak with a smaller weight. In other words, competition between different related magnetic states is favorable for superconductivity. In fact, it is doubly favorable, because it also suppresses the competing phase, i.e. SDW order.

References

- Superconductivity, science: britannica.com, Retrieved 27 May, 2019

- Superconductivity: chemicool.com, Retrieved 18 July, 2019

- Higher-temperature-superconductivity, science-superconductivity: britannica.com, Retrieved 12 January, 2019

- low-high-temperature-superconductivity: cesur.en.ankara.edu.tr, Retrieved 04 April, 2019

Models and Theories in Superconductivity

Superconductivity is a vast subject that comprises of several models and theories. The London equations, Ginzburg-Landau theory, Bardeen-Cooper-Schrieffer theory, Bean critical state model in superconductivity are some examples. This chapter discusses in detail these models and theories related to superconductivity.

The London Equations

The brothers F. and H. London wrote down two simple equations which conveniently incorporate the electrodynamic response of a superconductor. These equations describe the microscopic electric $(\mathbf{E})$ and magnetic $(\mathbf{h})$ fields inside a superconductor. Here $\mathbf{h}$ is the microscopic flux density, and B will be the macroscopic averaged flux density.

To derive the first London equation, think of the net force acting on the charge carrier in a normal metal:

$$\frac{d(m\mathbf{v})}{dt} = e\mathbf{E} - \frac{m\mathbf{v}}{\tau}$$

(Note that this is a LOCAL equation. It assumes that only the local electric field influences the drift velocity. As such, it requires the mean free path be less than the magnetic penetration depth, $\ell_{\mathrm{MFP}} < \lambda_{\mathrm{L}}$). Here v is the average or "drift" velocity of the charge carrier of charge e, m is its mass, E is the local electric field, and τ is a phenomenological scattering time for the carrier which describes how long it takes the scattering to bring the velocity of the carrier to zero. In a normal metal in steady state, the drift velocity achieves a constant value, meaning that the electric force and scattering forces balance, leading to:

$$\langle \mathbf{v} \rangle = e\mathbf{E}\tau / m$$

If there are n carriers per unit volume, the current density can be written as $J = ne\langle \mathrm{v} \rangle$, so

$$\mathbf{J} = \frac{ne^2\tau}{m}\mathbf{E}$$

which is Ohm's Law $(\mathbf{J} = \sigma\mathbf{E})$ with the conductivity $\sigma = \mathrm{ne}^2\tau / m$.

To model a superconductor, we shall suppose that there is a density of superconducting electrons, ns, and they do not have their velocities reduced to zero by means of scattering. From the above equation, this means that the electrons will accelerate in an applied electric field! $m\partial\vec{v}/\partial t = e\vec{E}$, giving rise to the first London equation:

$$\frac{\partial \mathbf{J}_s}{\partial t} = \frac{n_s e^2}{m} \mathbf{E}$$

or

$$\frac{\partial(\Lambda \mathbf{J}_s)}{\partial t} = \mathbf{E}$$

Strictly speaking, this equation only holds for ac currents and electric fields, since it predicts very large currents for large times at dc. The first London equation says that in order to create an alternating current (i.e.a non-zero $\partial \vec{J}_s / \partial t$) it is necessary to establish an electric field in the superconductor. This has implications for the finite-frequency losses in superconductors. If any un-paired electrons (quasiparticles) are around, they will be accelerated by the electric field and cause Ohmic dissipation. Hence a superconductor has a small but finite dissipation when illumi-nated with a finite frequency electromagnetic wave at temperatures above zero Kelvin. Supercon-ductors are only dissipation-less at zero frequency, or at finite frequency at zero temperature (for a fully-gapped superconductor).

We define a new quantity, Λ as,

$$\frac{\partial \mathbf{J}_s}{\partial t} = \frac{1}{\Lambda} \mathbf{E} = \frac{1}{\mu_0 \lambda_L^2} \mathbf{E}$$

where,

$$\Lambda = \mu_0 \lambda_L^2 = m/(n_s e^2).$$

The (London) magnetic penetration depth, λ_L. It is defined as,

$$\lambda_L = \sqrt{\frac{m}{\mu_0 n_s e^2}}$$

To get a deeper insight into the first London Equation and this new length scale, start with the Maxwell equation for the microscopic fields (Ampere's Law),

$$\nabla \times \vec{B} = \mu_0 \mathbf{J}_s + \mu_0 \frac{\partial \vec{D}}{\partial t}$$

and ignoring the displacement current (this is usually appropriate in superconductors because we often consider only frequencies $\omega < 2\Delta$), take the time derivative of both sides and use the first London equation, to obtain,

$$\nabla \times \frac{\partial \vec{B}}{\partial t} = \mu_0 \frac{\partial \mathbf{J}_s}{\partial t} = \frac{1}{\lambda_L^2} \mathbf{E}$$

Now take the curl of both sides,

$$\nabla \times \nabla \times \frac{\partial \vec{B}}{\partial t} = \frac{1}{\lambda_L^2} \nabla \times \mathbf{E}$$

The electric field curls around the time-varying magnetic field (Faraday's law),

$$\nabla \times \vec{E} = -\frac{\partial \vec{B}}{\partial t}$$

to get,

$$\nabla \times \nabla \times \frac{\partial \vec{B}}{\partial t} = -\frac{1}{\lambda_L^2} \frac{\partial \vec{B}}{\partial t}$$

Integrating both sides with respect to time yields,

$$\nabla \times \nabla \times \vec{B} + \frac{1}{\lambda_L^2} \vec{B} = 0.$$

Now use the vector identity,

$$\nabla \times \nabla \times \vec{B} = \nabla \left(\nabla \cdot \vec{B} \right) - \nabla^2 \vec{B}$$

And the fact that $\nabla \cdot \vec{B} = 0$ to arrive at,

$$\nabla^2 \vec{B} = \frac{1}{\lambda_L^2} \vec{B}$$

This equation admits solutions of the general form $B(x) = B_0 e^{\pm x/\lambda_L}$, so the London penetration depth λ_L represents the exponential screening length of the magnetic field in the superconductor. This length scale is also commonly referred to as the "magnetic penetration depth" for obvious reasons. This equation shows that the magnetic field is excluded from the bulk of a superconductor, and describes the Meissner effect.

A similar result can be derived for the electric field:

$$\nabla^2 \vec{E} = \frac{1}{\lambda_L^2} \vec{E},$$

showing that it is screened on the same length scale.

Ginzburg-Landau Theory

It turns out that for conventional (low-T_c) superconductors, mean field theory is an accurate description because fluctuations are tiny except very close to the transition temperature. This is not the case for high T_c superconductors.

Free Energy Expansion

For a complex order parameter Ψ the Landau expansion of the free energy for small $|\Psi|$ would be:

$$F\int\left[\alpha(T)|\Psi|^2+\frac{1}{2}\beta(T)|\Psi|^4+\gamma(T)|\nabla\Psi|^2\right]d^3x$$

For a charged superfluid we must add the coupling to the vector potential and also the magnetic energy, so that the full expression for a pair-superconductor is:

$$F\int\left[\alpha(T)|\Psi|^2+\frac{1}{2}\beta(T)|\Psi|^4+\gamma(T)\left|\left(\nabla+\frac{2ie}{\hbar c}\mathbf{A}\right)\Psi\right|^2+\frac{B^2}{8\pi}\right]d^3x.$$

Near the transition temperature T_c we can write $\alpha(T)\to a(T-T_c),\beta\to b$, and take a, b, γ to be independent of T. The free energy F must be minimized with respect to variations of Ψ and A.

$$\frac{\delta F}{\delta \mathbf{A}}=0=-\frac{2e\gamma}{\hbar c}i\left[\Psi^*\left(\nabla+\frac{2ie}{\hbar c}\mathbf{A}\right)\Psi-\Psi\left(\nabla-\frac{2ie}{\hbar c}\mathbf{A}\right)\Psi^*\right]+\frac{1}{4\pi}\nabla\times(\nabla\times\mathbf{A})$$

or,

$$\nabla\times\mathbf{B}=(4\pi/c)\mathbf{j}$$

with,

$$\mathbf{j}=-\frac{4e}{\hbar}\gamma|\Psi|^2\left(\nabla\phi\frac{2e}{\hbar c}\mathbf{A}\right).$$

These expressions give the London penetration depth:

$$\lambda^{-2}=32\pi\left(\frac{e^2}{\hbar^2c^2}\right)\gamma|\Psi|^2,$$

$$\gamma|\Psi|^2=n_s\frac{\hbar^2}{8m}.$$

Again only this product of parameters has real significance, but often the choice is made,

$$|\Psi|^2=\frac{1}{2}n_s,\quad \gamma\frac{\hbar^2}{4m},$$

and then,

$$\mathbf{j}=-\frac{e\hbar}{m}|\Psi|^2\left(\nabla\phi\frac{2e}{\hbar c}\mathbf{A}\right).$$

These results show that near T_c the superfluid density varies as $n_s \propto (1 - T/T_c)$.

Minimizing with respect to Ψ (or actually Ψ^*) gives:

$$\frac{\delta F}{\delta \Psi^*} = 0 = \alpha(T)\Psi + \beta(T)|\Psi|^2 \Psi - \gamma(T)\left(\nabla + \frac{2ie}{\hbar c}\mathbf{A}\right)^2 \Psi.$$

Using the convention $|\Psi|^2 = \frac{1}{2}n_s$, $\gamma\frac{\hbar^2}{4m}$, this can be written in a form that makes an analogy to the Schrodinger equation for particles of mass $2m$ apparent:

$$\frac{1}{4m}\left(-\hbar\nabla + \frac{2e}{c}\mathbf{A}\right)^2 \Psi + \alpha(T)\Psi + \beta(T)|\Psi|^2 \Psi = 0,$$

although this is a convenience for solving the equation, rather than anything deep, and purists don't like it because n_s is a stiffness constant and it is more natural to normalize the order parameter in terms of the strength of long range correlations, such as we did in the last lecture.

Near the transition temperature $\frac{\delta F}{\delta \Psi^*} = 0 = \alpha(T)\Psi + \beta(T)|\Psi|^2 \Psi - \gamma(T)\left(\nabla + \frac{2ie}{\hbar c}\mathbf{A}\right)^2 \Psi$ becomes:

$$-a(T_c - T)\Psi + b|\Psi|^2 \Psi - \gamma\left(\nabla + \frac{2ie}{\hbar c}\mathbf{A}\right)^2 \Psi = 0.$$

Correlation Length

The solution to equation above for the uniform state is:

$$|\Psi|^2 = a(T_c - T)/b$$

giving the usual square root growth of the order parameter for $T < T_c$ found in mean field theories. The corresponding free energy density for $T < T_c$ is,

$$f = \frac{F}{V} = -\frac{a^2(T_c - T)^2}{2b},$$

giving the jump in the specific heat at T_c by two differentiations with respect to temperature. From our calculation of the zero temperature energy we would guess,

$$F(T \simeq T_c) \sim -N\frac{(k_B T_c)^2}{\varepsilon_F}\left(1 - \frac{T}{T_c}\right)^2$$

so that,

$$\frac{a^2}{b} \sim \frac{nk_B^2}{\varepsilon_F}.$$

For spatial variations of the order parameter equation,

$$-a(T_c - T)\Psi + b|\Psi|^2 \Psi - \gamma\left(\nabla + \frac{2ie}{\hbar c}\mathbf{A}\right)^2 \Psi = 0$$

yields the length scale ξ given by,

$$\xi^2 = \frac{\gamma}{\alpha} = \frac{\gamma}{a}(T_c - T)^{-1},$$

or $\xi = \xi_0(1 - T/T_c)^{-1/2}$ with the temperature independent length scale $\xi_0 = \gamma/aT_c$. On the other hand from $\gamma|\Psi|^2 = n_s \dfrac{\hbar^2}{8m}$ we estimate (supposing $n_s(T \to 0) \sim n$),

$$\gamma\frac{a}{b}T_c \sim n\frac{\hbar^2}{2m},$$

so that,

$$\xi_0^2 \sim \frac{\hbar^2}{2M\varepsilon_F}\left(\frac{\varepsilon_F}{k_B T_c}\right)^2.$$

The correlation length is a therefore a factor of $\varepsilon_F/k_B T_c$ larger than $\left(\hbar^2 2m\varepsilon_F\right)^{1/2}$ which is of order the interparticle spacing. This can be traced to the fact that the stiffness constant is determined by the total density of electrons, whereas the energy coefficients a, b are given by the fraction of particles corresponding to the energy band of width about $k_B T_c$ around the Fermi surface that is affected by the pairing.

The Ginzburg criterion for the temperature T_G near T_c when fluctuations become important can be estimated as $f\xi^3 \sim k_B T_c$. With the above results this is estimated as $1 - T_G/T_c = t_G$ given by:

$$\frac{n(k_B T_c)^2}{\varepsilon_F} t_G^{1/2} \left[\frac{\hbar^2}{2m\varepsilon_F}\left(\frac{\varepsilon_F}{k_B T_c}\right)^2\right]^{3/2} \sim k_B T_c$$

i.e. $t_G \sim (k_B T_c/\varepsilon_F)^4$. This is very small for conventional (not high-T_c) superconductors, so that fluctuation corrections and the critical region near T_c are usually immeasurable.

Behavior in a Magnetic Field

Dimensionless Equation

In a constant imposed magnetic field $\mathbf{H}$ (i.e. the field due to external current sources) the appropriate free energy to minimize is:

$$G_H = F - \mathbf{B}(\mathbf{x}) \cdot \mathbf{H}/\pi.$$

In the normal state $\mathbf{B}(x) = \mathbf{H}$, whereas in the bulk superconducting state $\mathbf{B} = 0$. Thus the simplest idea would be that superconductivity is killed at a thermodynamic critical field H_c given by:

$$\frac{H_c^2}{8\pi} = \frac{\alpha^2}{2\beta}.$$

For fields larger than this the system becomes normal. This would be disappointing, since for typical superconductors this critical field would be only a few hundred gauss even at low temperatures. Luckily, the behavior in a field is more complicated than this, and in some materials superconductivity persists up to much higher fields, a vital result for the technological applications that are common today.

To proceed it is useful to simplify the Ginzburg-Landau equation by introducing dimensionless units: Measure lengths in units of the London penetration depth λ, magnetic fields in units of $\sqrt{2}H_c$, the order parameter in units of $|\alpha|/\beta$, and the energy density in units of $H_c^2/4\pi$, i.e. define,

$$\mathbf{x'} = \mathbf{x}/\lambda,$$

$$\mathbf{A'} = \mathbf{A}/\sqrt{2}H_c\lambda,$$

$$\mathbf{B'} = \mathbf{B}/\sqrt{2}H_c,$$

$$f' = 4\pi f/H_c^2,$$

$$\Psi' = \Psi/(|\alpha|/\beta),$$

Then (dropping the primes) the free energy density is:

$$f = -|\Psi|^2 + \frac{1}{2}|\Psi|^4 + \left|\left(-ik^{-1}\nabla + \mathbf{A}\right)\Psi\right|^2 + B^2$$

with,

$$\kappa = \frac{\lambda}{\xi} = \frac{1}{4}\left(\frac{\hbar c}{e\gamma}\right)\left(\frac{\beta}{2\pi}\right)^{1/2}.$$

Although λ and ξ both depend strongly on temperature, this mainly cancels out in the ratio, and κ is roughly temperature independent. It is the key parameter in determining the nature of the behavior in a magnetic field. Since λ and ξ derive from quite different physics, κ varies from small to large values in different materials.

In the dimensionless units the Ginzburg-Landau equation is:

$$\left(-i\kappa^{-1}\nabla + \mathbf{A}\right)^2 \Psi - \Psi + |\Psi|^2\Psi = 0,$$

and the current current equation is,

$$\nabla \times \mathbf{B} = \frac{1}{2}\left[\Psi^*\left(-i\kappa^{-1}\nabla + \mathbf{A}\right)\Psi - \Psi\left(i\kappa^{-1}\nabla + \mathbf{A}\right)\Psi^*\right]$$

The flux quantization condition in the scaled units is given by identifying the current from above equation:

$$\mathbf{j} \propto |\Psi|^2 \left(\kappa^{-1}\nabla\phi + \mathbf{A}\right)$$

so that in the bulk of a superconductor where $\mathbf{j} = 0$.

$$\int \mathbf{B}\cdot d\mathbf{S} = \oint \mathbf{A}\cdot d\mathbf{l} = \text{integer} \times 2\pi\kappa^{-1}.$$

Thus the flux quantum is $2\pi\kappa^{-1}$ in the scaled units.

Surface Energy

If we set the external field to the thermodynamic critical field H_c the normal and superfluid states have the same free energy and can be in contact, and the question of the surface energy between them can be addressed. We choose the direction normal to the surface to be $\mathbf{z}$ and $\mathbf{A} = A(z)\hat{\mathbf{x}}, \mathbf{B} = B(z)\hat{\mathbf{y}}$ the governing equations are:

$$-\kappa^{-2}\frac{d^2\Psi}{dz^2} + \Psi - |\Psi|^2\Psi + A^2\Psi = 0.$$

$$\frac{d^2 A}{dz^2} - |\Psi|^2 A = 0.$$

Where it is consistent to assume only z dependence and no phase variation. The surface free energy is the difference of G_H from the value with all superconductor or normal state. In the scaled units and at $H = H_c$ this is:

$$\sum \int_{-\infty}^{\infty} dz \left[-|\Psi|^2 + \frac{1}{2}|\Psi|^4 + \left|\left(-i\kappa^{-1}\frac{d}{dz} + A\right)\Psi\right|^2 + \left(B - \frac{1}{\sqrt{2}}\right)^2\right].$$

If $-\kappa^{-2}\dfrac{d^2\Psi}{dz^2} + \Psi - |\Psi|^2\Psi + A^2\Psi = 0$ is multiplied by Ψ^*, integrated over z, and then we integrate

by parts, $\sum$ can be simplified to,

$$\sum \int_{-\infty}^{\infty} dz \left[-\frac{1}{2}|\Psi|^4 + \left(B - \frac{1}{\sqrt{2}}\right)^2\right].$$

Note that the integrand is zero in the superconductor ($\Psi = 1, B = 0$) and the normal state ($\Psi = 0, B = 1/\sqrt{2}$). Thus $\sum$ gets contributions just from the interface region, as makes sense.

Equations above must be solved numerically, but we can deduce the main results without a full solution. For κ large, $\xi << \lambda$, the field penetrates a large distance into the superconductor compared to the thickness of the interface. Thus in the interface region the integrand in equation above is negative, and the surface energy is negative. On the other hand for small $\kappa, \lambda >> \xi$, the field cannot penetrate even the small Ψ region before the superconducting state is established, the integrand is positive in the interface region, and so the surface tension is negative. The dividing value of κ is given by $\sum = 0$, which means $|\Psi(z)|^2 (B(z) - 1/\sqrt{2})$. It can be shown that this satisfies

$$-\kappa^{-2} \frac{d^2 \Psi}{dz^2} + \Psi - |\Psi|^2 \Psi + A^2 \Psi = 0 \text{ for } \kappa = 1/\sqrt{2}.$$ This divides superconductors into two classes:

Type I: $\kappa < 1/\sqrt{2}$: positive surface tension.

Type II: $\kappa > 1/\sqrt{2}$: negative surface tension.

For type I superconductors the thermodynamic critical field H_c does indeed give the boundary between superconducting and normal states. For type II superconductors however the negative surface energy favors a mixed state of regions of normal and superconductor intermixed on length scales of order ξ or λ for fields $H_{c1} < H < H_c$ where the negative surface energy favors the invasion of normal regions with magnetic field into the superconductor, and for $H_c < H < H_{c2}$ where the negative surface energy further stabilizes the superconducting state.

Lower Critical Field

The lower critical field H_{c1} for a type II superconductor occurs when the energy for a single quantized flux line or vortex becomes negative. There is a positive contribution to the energy density from the phase gradient $\nabla \sim$ at a distance r from the vortex core. From

$$f = -|\Psi|^2 + \frac{1}{2}|\Psi|^4 + |(-ik^{-1}\nabla + A)\Psi|^2 + B^2$$ this gives an energy density per unit length of line of

order $\kappa^{-2} r^{-2}$ in the scaled units (taking $|\Psi| \sim 1$), over distances from ξ to λ the London penetration depth. The integral gives the energy cost:

$$\Delta G_+ \sim \kappa^{-2} \ln(\kappa).$$

In addition there is a negative contribution from the single quantum of flux in the magnetic field H_{c1}. this decrease in magnetic energy (flux × external field) per length of line is of order:

$$\Delta G_- \sim \kappa^{-1} H_{c1}.$$

Thus the lower critical field is of order:

$$H_{c1} \sim \kappa^{-1} \ln \kappa.$$

The calculation can be done essentially exactly for large κ when the local London equation can be used, and the result is actually $H_{c1} = (2\kappa^{-1})(\ln C\kappa)$ with C a number of order unity. Near H_{c1} the density of vortices d^{-2} is determined by balancing the repulsive energy of the interacting vortices against the magnetic energy gained. Since the interaction is exponential in the separation $\sim e^{-d}$, whereas the magnetic energy gained per vortex is proportional to $(H - H_{c1})$ there is a rapid increase in the density of vortices $\sim [\ln(H - H_{c1})]^{-2}$. The average magnetic field B scales in the same way. We would expect the repulsive interaction to lead to a lattice structure, perhaps a close packed triangular lattice.

Intermediate Fields

As the external field is increased, the density of vortices increases, and the average magnetic field over the superconductor grows, initially rapidly since the flux lines interact weakly. When the separation d becomes comparable with the penetration depth λ the supercurrent and field regions begin to extend over the whole superconductor, the flux lines interact more strongly, and the growth of B with H is slower. When $d \sim \xi$ the normal cores of the flux lines overlap, and the state becomes normal. In scaled units this is $d \sim \kappa^{-1}$ and the magnetic field is one flux quantum per d^2 or of order $\kappa^{-1} / \kappa^{-2} \sim \kappa$. In the normal state $B = H$ so this gives us the estimate of the upper critical field $H_{c2} = \kappa$ (i.e. $\sqrt{2}\kappa H_c$ in unsclaed units). We will see this is actually the exact result.

Upper Critical Field

At the upper critical field the vortex cores overlap suppressing the order parameter close to zero. This means we can linearize $\left(-i\kappa^{-1}\nabla + \mathbf{A}\right)^2 \Psi - \Psi + |\Psi|^2 \Psi = 0$, in Ψ to give:

$$\left(-i\kappa^{-1}\nabla + \mathbf{A}\right)^2 \Psi - \Psi = 0.$$

Neglect the feedback of the supercurrents on the magnetic field so that for an applied field $H\hat{\mathbf{z}}$ we can take $\mathbf{A} = Hx\hat{\mathbf{y}}$ and then:

$$\left[-\kappa^{-2}\nabla^2 - 2i\kappa^{-1}Hx\frac{\partial}{\partial y} + H^2 x^2\right]\Psi = \Psi.$$

This is the same as Schrodinger's equation for particles in a constant magnetic field, giving Landau levels. Note that although the physical problem is symmetric in $x \to y$, by choice of gauge we have formulated the problem in a way that does not respect this symmetry. This is also what is usually done for the Landau level problem. Assuming a z-independent solution, we can write:

$$\Psi = e^{i\kappa_y y}u(x),$$

with u(x) satisfying,

$$-u'''' + (H\kappa x - k_y)^2 u = \kappa^2 u.$$

This is the same as Schrodinger's equation for a harmonic oscillator about $x_0 = k_y / \kappa H$ and there are bounded solutions for:

$$\kappa = (2n+1)H.$$

The largest critical field corresponds to $n = 1$ giving,

$$H_{c2} = \kappa.$$

The corresponding eigenfunction is,

$$\Psi = e^{ik_y y} e^{-\kappa^2 (x - \kappa^2 k_y)^2 / 2}$$

If we suppose a periodic solution in the y direction with period L_y then the physical solution can be any linear combination of the solutions with $k_y = n \times 2\pi / L_y$,

$$\Psi = \sum_n C_n e^{i(2n\pi / L_y) y} e^{-\kappa^2 [x - \kappa^{-2}(2n\pi / L_y)]^2 / 2} .$$

First consider the case $C_n = C$. Then increasing x by $\kappa^{-2}(2\pi / L_y)$ simply corresponds to a phase change of Ψ by $2\pi y / L_y$ so that $|\Psi|^2$ is periodic in x with period L_x satisfying:

$$L_x L_y = 2\pi \kappa^{-2}.$$

The individual values of L_x and L_y (and the competition with other solutions with different choices of C_n) is not given by this linear calculation. Keeping the lowest order nonlinear terms in $|\Psi|^2$ in both:

$$\left(-i\kappa^{-1}\nabla + \mathbf{A}\right)^2 \Psi - \Psi + |\Psi|^2 \Psi = 0,$$

and,

$$\nabla \times \mathbf{B} = \frac{1}{2}\left[\Psi^* \left(-i\kappa^{-1}\nabla + \mathbf{A}\right)\Psi - \Psi\left(i\kappa^{-1}\nabla + \mathbf{A}\right)\Psi^*\right]$$

is quite involved. Not surprisingly, for the case of $C_n = C$ the free energy is minimized for $L_x = L_y$ which actually gives a structure with square symmetry (not immediately obvious, since our representation of x and y dependence is quite different). The structure consists of a square lattice of points where $|\Psi|$ goes to zero, about which the phase winds by 2π— the structure of a rudimentary lattice of vortices. Actually it turns out that the lowest free energy solution is for a two parameter solution $C_{2n} = C_0, C_{2n+1} = C_1$ with $C_1 = iC_0$ and a ratio of L_x / L_y which turns out to correspond to a triangular lattice. Since the same structure is expected in the dilute flux line limit near H_{c1} this flux lattice structure is expected over the whole range of fields $H_{c1} < H < H_{c2}$. The lattice structure is dramatically confirmed by experiments where the points of large field on the surface where flux lines emerge are decorated with small magnetic particles.

The flux-lattice structure in type II superconductors is enormously important in technological applications, allowing fields enhanced by the factor κ to be sustained. Unlike the case of superfluids, where vortex lines are immediately mobile and dissipative, flux lines are typically pinned to impurities in the lattice (κ tends to be large in "dirty superconductors"). In high-T_c superconductors the thermal fluctuations of the flux lines becomes important, and indeed the superconducting transition in a magnetic field must be thought of as the melting of the flux line lattice, since in a disordered flux-line state there is no long range phase order.

Bardeen-Cooper-Schrieffer Theory

The BCS theory or Bardeen-Cooper-Schrieffer theory is the theory of superconductivity developed by John Bardeen, Leon N Cooper and John R. Schrieffer.

Electron Correlations that Produce Superconductivity

The superconducting phase exhibits correlations absent in the normal metal. It is believed that any attractive interaction between fermions in a many-fermion system can produce a superconducting-like state and this is thought to be the case, in addition to metals, in nuclei, neutron stars and He[3]. Thus the BCS theory focuses on the consequences of an attractive two-body interaction without enquiring too much further about its origin.

The fundamental difference between the superconducting and normal ground state wave functions is produced when the large degeneracy of single-particle electron levels in the normal state is removed. If one visualizes the Hamiltonian matrix that results from an attractive two-body interaction in the basis of normal metal configurations, one finds sub-matrices in which all single-particle states except for one pair of electrons remain unchanged. These two electrons can scatter via their interaction to all states of the same total momentum. Such pair may be thought to "wend its way" over all states unoccupied by other electrons. It is known as a Cooper pair and plays the central role in BCS.

Since every such state is connected to every other, one is presented with submatrices of the entire Hamiltonian corresponding to an M-dimensional space of two-particle excited states on top of the Fermi sea. The origin of Cooper pairs can be understood from the following simplified example: all off-diagonal elements are set equal to $-V$ and all diagonal terms are set equal to 0 as though all the initial electron levels were completely degenerate,

$$
\begin{pmatrix}
0 & -V & -V & -V & . & . & . & . & . & . & & . & -V \\
-V & 0 & -V & . & . & . & . & . & . & & . & & . \\
-V & -V & 0 & -V & . & . & & . & . & & & & . \\
. & . & . & . & . & . & & & & & & & . \\
. & . & & . & & . & & & & & & & . \\
. & . & & & . & & & . & . & . & . & & . \\
. & & & . & & & . & & . & . & . & & . \\
. & & & & & . & & . & & & & & . \\
. & . & . & & & . & . & . & . & . & & \\
. & . & . & & . & . & . & . & -V & 0 & -V \\
-V & -V & . & . & . & . & . & . & -V & -V & 0
\end{pmatrix}
$$

Diagonalizing this matrix results in an energy level structure with $M-1$ levels with energy V and one level with the energy $E=-(M-1)V$. The latter level is a superposition of all original levels.

The number of levels M is proportional to the volume of the system while the scattering matrix element V is inversely proportional to the volume. Hence, E is independent of the volume. In other words, the removal of the degeneracy produces a single level separated from the others by a volume independent gap.

To incorporate this into a solution of the full Hamiltonian, one must devise a technique by which all electron pairs can scatter while obeying the exclusion principle. In the BCS wave function that accomplishes this, an inspired guess of Robert Schrieffer, each pair gains an energy due to the removal of the degeneracy as above and one obtains the maximum correlation of the entire wave function if all pairs have the same total momentum. For a combination of statistical and dynamical reasons, this gives a coherence to the wave function with a preference for momentum zero, singlet spin correlations. Formation of the superconducting state is a phase transition, and is characterized by an order parameter. Gor'kov showed that the superconducting symmetry parameter is proportional to the wave function of the Cooper pair.

Ground State

In the simplest model only singlet zero-momentum pairs interact and scatter. They can be conveniently described by the pair operators:

$$b_k = c_{-k\downarrow c k\uparrow}$$

$$b_k^\dagger = c_{k\uparrow}^\dagger c_{-k\downarrow}^\dagger,$$

where $c_{k\uparrow,\downarrow}$ are fermion annihilation operators, the arrow shows the spin projection. We expect that in a good approximation the wave function can be constructed from the ground state of a non-interacting electron gas entirely with combinations of the pair operators. Using this one extracts from the full Hamiltonian the so-called reduced Hamiltonian,

$$H_{reduced} = \sum_{k<k_f} 2\,|\,\epsilon_k\,|\,b_k b_k^\dagger + 2 \sum_{k>k_f} \epsilon_k b_k^\dagger b_k + \sum_{k,q} V_{kp} b_k^\dagger b_p,$$

where k_f is the Fermi momentum, ϵk are single electron energy levels (the energy is zero at the Fermi level) and V_{kp} is the scattering matrix element between the pair states k and p. The reduced Hamiltonian has correct matrix elements in the subspace built with pair operators.

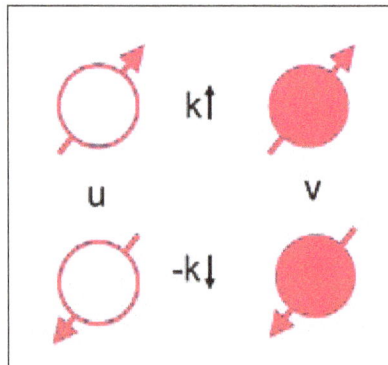

The ground state of the superconductor is a linear superposition
of states in which pairs $(k\uparrow,-k\downarrow)$ are occupied or unoccupied.

The ground state of the superconductor is a linear combination of pair states in which the pairs $(k\uparrow, -k\downarrow)$ are occupied (state O_k) or unoccupied (state $O_{(k)}$) as indicated in figure:

$$\psi_0 = u_k O_{(k)} + v_k O_k.$$

The probability amplitude that the pair state is (is not) occupied is then v_k (u_k). Normalization requires that $|u|^2 + |v|^2 = 1$. The phase of the ground state may be chosen so that with no loss of generality u_k is real. We can then write:

$$u = \sqrt{1-h}; \, v = \sqrt{h}\exp(i\phi),$$

where $0 \le h \le 1$.

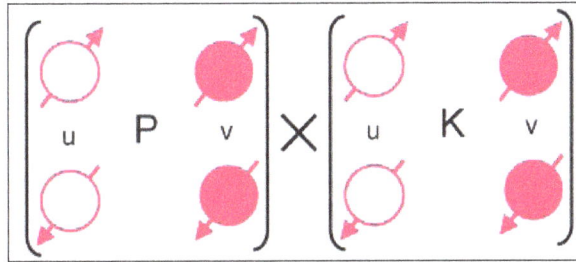

A decomposition of the ground state of the superconductor into
states in which the pair states k and p are either occupied or unoccupied.

A further decomposition of the ground state wave function in which pair states k and p are occupied or unoccupied is:

$$\psi_0 = (u_k O_{(k)} + v_k O_k)(u_p O_{(p)} + v_p O_p).$$

This is a Hartree-like approximation in the probability amplitudes for the occupation of pair states. The decomposition procedure should be continued to include all k, so that the complete BCS ansatz for the ground state is given by:

$$\psi_0 = \prod_k (u_k O_{(k)} + v_k O_k).$$

This wave function does not have a fixed number of particles N, but the relative fluctuations of N are of the order $1/\sqrt{N}$ in the thermodynamic limit, so negligible for macroscopic samples. Minimizing the energy:

$$W = \left\langle \psi_0 \left| H_{\text{reduced}} \right| \psi_0 \right\rangle$$

with respect to the variational parameters h_k and ϕ_k gives,

$$h = (1 - \epsilon / E)/2,$$

where,

$$E(k) = (\epsilon_k^2 + |\Delta(k)|^2)^{1/2}$$

and $\Delta(k)$ satisfies the equation,

$$\Delta(k) = -\frac{1}{2}\sum_q V_{kq} \frac{\Delta(q)}{E(q)}.$$

If a non-zero solution of this integral equation exists, $W < 0$, then the normal Fermi sea is unstable under the formation of Cooper pairs. In the wave function that results there are strong correlations between pairs of electrons with opposite spin and zero total momentum. These correlations are built from normal excitations near the Fermi surface and extend over spatial distances typically of the order of 10^{-4} cm. They can be constructed due to the large wave numbers available because of the exclusion principle. Thus, with a small additional expenditure of kinetic energy there can be a greater gain in the potential energy term.

Single-particle Excitations

In considering the excited states of the superconductor it is useful to make a distinction between single-particle and collective excitations: it is the single-particle excitation spectrum whose alteration is responsible for superfluid properties. For the superconductor, excited (quasiparticle) states can be defined in one-to-one correspondence with the excitations of the normal metal. One finds, for example, that the expectation value of H_{reduced} for the excitation in figure is given by:

$$E = \sqrt{\epsilon_k^2 + |\Delta|^2}.$$

In contrast to normal systems, for the superconductor even as ϵ goes to zero, the excitation energy E remains larger than zero, its lowest positive value is $|\Delta|$. One can therefore produce single particle excitations from the superconducting ground state only with the expenditure of a small but finite amount of energy. This is called the energy gap (in gapless superconductors $\Delta(k)$ is zero at special directions of the momentum). In the ideal superconductor, the energy gap appears because not a single pair can be broken nor a single element of the phase space be removed without a finite energy cost. If a single pair is broken, one loses its correlation energy W; if one removes an element of phase space from the system, the number of possible transitions of all pairs is reduced resulting in both cases in an increase in the energy which does not go to zero as the volume of the system increases.

The excitation spectrum of the superconductor can be conveniently treated by introducing a linear combination of the fermion creation and annihilation operators. This is known as the Valatin-Bogoliubov transformation:

$$\gamma_{k\uparrow}^\dagger = u_k c_{k,\uparrow}^\dagger - v_k^* c_{-k,\downarrow};$$

$$\gamma_{k\downarrow}^\dagger = v_k^* c_{k,\uparrow} + u_k c_{-k,\downarrow}^\dagger.$$

It follows that,

$$\gamma_{k\sigma}\psi_0 = 0$$

so that the $\gamma_{k\sigma}$ play the role of annihilation operators, while $\gamma_{k\sigma}^{\dagger}$ create excitations:

$$\gamma_{ki}^{\dagger} \cdots \gamma_{qj}^{\dagger} \psi_0 = \psi_{ki}, \ldots, _{qj}.$$

The γ operators satisfy Fermi anti-commutation relations so that they produce a complete set of fermionic excitations in one-to-one correspondence with the excitations of a normal metal. In the ground state of the superconductor all the electrons are in singlet-pair correlated states of zero total momentum. In an m-electron excited state the excited electrons are in a 'quasipartilce' states, very similar to Fermi liquid excitations and not strongly correlated with any other electrons. The other electrons are still correlated much as they were in the ground state. The excited electrons behave in a manner similar to normal electrons; they can be easily scattered or excited further. But the background electrons – those which remain correlated – retain their special behavior; they are difficult to scatter or excite.

Thus, one can identify two almost independent fluids. The correlated portion of the wave function shows the resistance to change and the very small specific heat characteristic of superfluid, while the excitations behave very much like normal electrons, displaying an almost normal specific heat and resistance. When a steady electric field is applied to the metal, the superfluid electrons short out the normal ones, but with higher frequency fields the resistive properties of the excited electrons can be observed.

Thermodynamic Properties of the Ideal Superconductor

The thermodynamic properties of the superconductor follow from the excitation spectrum . The free energy of the superconductor is given by:

$$F[h, \phi, f] = W(T) - TS,$$

where T is the temperature and S is the entropy; $f(k)$ is the probability that the state k is occupied by a quasiparticle (a superconducting Fermi function). The entropy of the system comes entirely from the excitations as the correlated portion of the wave-function is not degenerate. Since a portion of the phase space is occupied by excitations at finite temperatures, making it unavailable for the transitions of bound pairs, the energy is a function of the temperature $W(T)$. As T increases, $|W(T)|$ and at the same time Δ decrease until the critical temperature is reached and the system reverts to the normal phase.

The minimization of the free energy with respect to the variational parameters h and f gives the Fermi-Dirac expression for f:

$$f = \frac{1}{1 + \exp(E / kT),}$$

where E is given by $E(k) = (\epsilon_k^2 + |\Delta(k)|^2)^{1/2}$ and the gap $\Delta|$ satisfies the fundamental integral equation of the theory:

$$\Delta_k(T) = -\frac{1}{2} \sum_q V_k q \frac{\Delta_q(T)}{E_q(T)} \tanh \frac{E_q(T)}{2kT}.$$

This equation has a nonzero solution for the gap $\Delta(T)$ below a critical temperature T_c. At $T < T_c$ the properties of the system are qualitatively different from the normal metal.

In the simplified model used in the original formulation of the BCS theory, the scattering potential was approximated as:

$$V_{kq} = -V, |\epsilon| < \hbar\omega_D$$

$$V_{kq} = 0, |\epsilon| > \hbar\omega_D$$

where ω_D is the Debye frequency of the material. The energy-dependent density of states was replaced by its value at the Fermi surface $N(0)$. In this approximation the second-order transition temperature in zero magnetic field:

$$kT_c = 1.14\hbar\omega_D \exp(-1/N(0)V).$$

At zero temperature the gap,

$$\Delta(0) = 1.76kT_c.$$

The Cooper pair size can be estimated as $\xi = hv_F/[\pi\Delta]$, where v_F is the Fermi velocity in the normal state. The gap approaches zero as:

$$\Delta(T) = 1.74\Delta(0)\sqrt{1-T/T_c}$$

near the critical temperature. The specific heat is discontinuous at the transition point. Right below the transition it is 2.43 times greater than immediately above it. At low temperatures the specific heat is exponentially small as a function of the inverse temperature.

Microscopic Interference Effects

In its interaction with external perturbations the superconductor displays remarkable interference effects which result form the paired nature of the wave function and are not at all present in similar normal metal interactions. Neither would they be present in any ordinary two-fluid model of superconductivity. This 'coherence effects' are in a sense manifestations of interference in spin and momentum space on a microscopic scale, analogous to the macroscopic quantum effects due to interference in ordinary space. They depend on the behavior under time reversal of the perturbing fields.

Near the transition temperature these coherence effects produce quite dramatic contrasts in the behavior of coefficients which measure interactions with the conduction electrons. Historically, the comparison with theory of the behavior of the relaxation rate of nuclear spins and the attenuation of longitudinal ultrasonic waves in clean samples as the temperature is decreased through T_c provided an early test of the detailed structure of the theory.

The attenuation of longitudinal acoustic waves due to their interaction with the conduction electrons in a metal undergoes a very rapid drop as the temperature drops below T_c. Since the scattering of phonons from 'normal' electrons is responsible for most of acoustic attenuation, a drop is to be expected both in BCS in a two-fluid model but the rapidity of the decrease is difficult to reconcile with theoretical estimates within a two-fluid model.

The rate of relaxation of nuclear spins was measured by Hebel and Slichter in zero magnetic field in aluminum $(T_c = 1.18K)$ from 0.92K to 4.2K just at the time of the development of the BCS theory in 1957, figure. The dominant relaxation mechanism is provided by interaction with the conduction electrons so that one would expect, on the basis of a two-fluid model, that this rate should decrease below the transition temperature due to the diminishing density of 'normal' electrons. The experimental results however show just the reverse. The relaxation rate does not drop but increases by a factor of more than two just below the transition temperature in agreement with BCS but contrary to the predictions of other approaches to the theory of superconductivity.

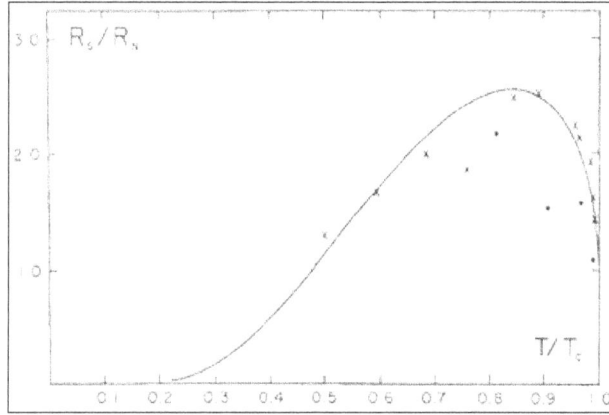

Comparison of observed nuclear spin relaxation rate with theory. The circles represent experimental data of Hebel and Slichter, the crosses data by Anderson and Redfield.

To understand how such effects come about in theory, one needs to consider the transition probability per unit time of a process involving electron transitions from the excited electron state k to the state p with the emission or absorption of energy from the interacting field. What is to be calculated is the rate of transition between an initial state $\langle i|$ and a final state $\langle f|$ with an absorption or emission of the energy $\hbar\omega_{|\vec{k}-\vec{p}|}$ (a phonon for example in the interaction of sound waves with the superconductor). All of this properly summed over final states and averaged with statistical factors over initial states may be written:

$$\Omega = \frac{2\pi}{\hbar} \frac{\sum_{i,f} \exp(-W_i / kT)\left|\langle f|H_{int}|i\rangle\right|^2 \delta(W_i - W_f)}{\sum_i \exp(-W_i / kT)},$$

where W_i are the energies in different states and H_{int} is the interaction Hamiltonian. The interaction Hamiltonian can typically be represented as,

$$H_{int} = \sum_{KP} B_{PK} c_P^\dagger cK,$$

where the operator B is the electronic part of the matrix element between the full final and initial states $\langle f|H_{int}|i\rangle = m_{fi}\langle f|B|i\rangle$, the indices K,P contain both the momentum k,p and spin; the spins of the states K and $-K$ are opposite.

In the normal system, scattering from single-particle electron states K to P is independent of scattering from −P to −K. But the superconducting states are linear combinations of (K,−K) occupied and unoccupied. Because of this, states with excitations $k\uparrow, p\uparrow$ are connected not only by $c^{\dagger}_{p\uparrow}c_{k\uparrow}$ but also by $c^{\dagger}_{-p\downarrow}c_{-k\downarrow}$: if the state |f⟩ contains the single-particle excitation $p\uparrow$ while the state |i⟩ contains $k\uparrow$, as a result of the superposition of occupied and unoccupied pair states in the coherent part of the wave function, these are connected not only by $B_{PK}c^{\dagger}_{P}c_{K}$ but also by $B_{-K-P}c^{\dagger}_{-K}c_{-P}$.

Many operators B (e.g., the electric current, or the charge density operator) have a well-defined behavior under the operation of time reversal so that:

$$B_{PK} = \pm B_{-K-P} = B_{p,k},$$

where the last expression does not contain spin indices. Then B becomes,

$$B = \sum_{kp} B_{pk}(c^{\dagger}_{p\uparrow}c_{k\uparrow} \pm c^{\dagger}_{-k\downarrow}c_{-p\downarrow}),$$

where the upper (lower) sign results for operators even (odd) under time reversal. As a result the matrix element squared $\left|\langle f|B|i\rangle\right|$ contains terms of the form $\left|B_{pk}\right|^{2}\left|u_{p}u_{k} \mp v_{p}v^{*}_{k}\right|^{2}$.

Applied to processes involving the emission or absorption of boson quanta such as phonons or photons, the squared matrix element above is averaged with the appropriate statistical factors over initial and summed over final states; subtracting emission from absorption probability per unit time, one typically obtains:

$$a = \frac{4\pi}{\hbar}\left|m\right|^{2}\sum_{kp}\left|u_{p}u_{k} \mp v_{p}v^{*}_{k}\right|^{2}(f_{p}-f_{k})\delta(E_{p}-E_{k}-\hbar\omega_{|\vec{k}-\vec{p}|}),$$

where f_{k} is the occupation probability for an excitation with the momentum k.

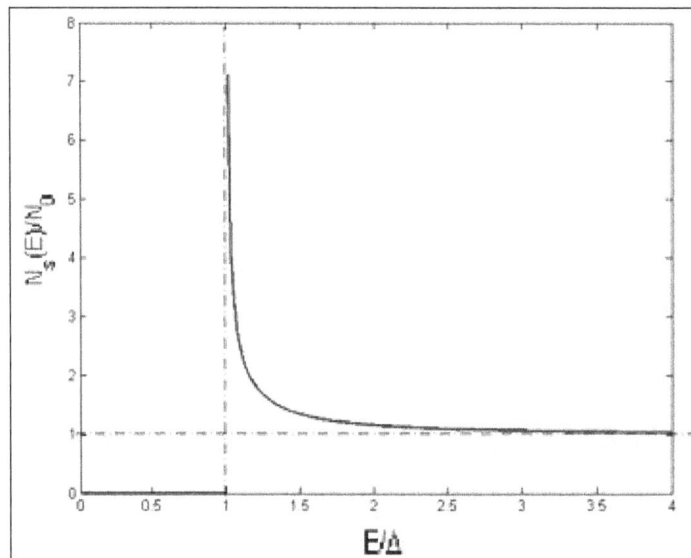

Ratio of superconducting to normal density as a function of E/Δ.

For the ideal superconductor there is isotropy around the Fermi surface and symmetry between particles and holes; therefore sums of the form $\sum_k$ can be converted to integrals over the superconducting excitation energy,

$$E : \sum_k \to 2N(0)\int_\Delta^\infty \sqrt{E/E2 - \Delta^2}\, dE,$$

where $N_s(E) = N(0)E/\sqrt{E^2 - \Delta^2}$ is the density of excitations in the superconductor.

The shape of Figure shows that contrary to intuitive expectations, the onset of superconductivity might enhance rather than diminish electronic transitions. However, the coherence factors:

$$|u_p u_k \mp v_p v_k^*|^2 = [1 + (\epsilon(k)\epsilon(p) \mp \Delta^2)/E(k)E(p)]/2$$

completely negate the effect of the increased density of state in the case of the operators even under time reversal. For the operators odd under time reversal the effect of the increase of the density of states is not cancelled and can be observed in the increase of the rate of the corresponding process. The current and spin are odd under time reversal while the charge density is even. The first two operators are responsible for electromagnetic interaction and nuclear spin relaxation interaction while the latter is relevant for the electron-phonon interaction which shows strikingly different effects.

Ultrasonic attenuation in the ideal pure superconductor for $ql \gg 1$ (the product of the phonon wave number and the electron mean free path) depends in a fundamental way on the absorption and emission of phonons. Since the matrix elements have a very weak dependence on changes near the Fermi surface in occupation numbers other than k or p that occur in the normal to superconducting transition, calculations within the quasi-particle model can be compared in a very direct manner with similar calculations for the normal metal, as B_{pk} is the same in both states. The ratio of the attenuation in the normal and superconducting states reduces to:

$$\frac{a_s}{a_n} = \frac{2}{1 + \exp(\Delta(T)/kT)}.$$

This prediction was used by Morse and Bohm for a direct experimental determination of the variation of Δ with T and obtained an excellent agreement with the BCS theory.

On the other hand, the ratio of the nuclear spin relaxation rates in superconducting and normal states in the same sample can be estimated as:

$$\frac{R_s}{R_n} \sim \int N_s^2(E)\,dE / \int N_n^2(E)\,dE,$$

where $N_s(E)$ and $N_n(E)$ stay for the normal and superconducting densities of states. Taken literally, in fact, this expression diverges logarithmically at the lower limit due to the infinite density of states. When the Zeeman energy difference between the spin up and spin down states is included, the integral is no longer divergent but the integrand is much too large. Quantitative agreement between theory and experiment was obtained by Fibich by including the effect of thermal phonons.

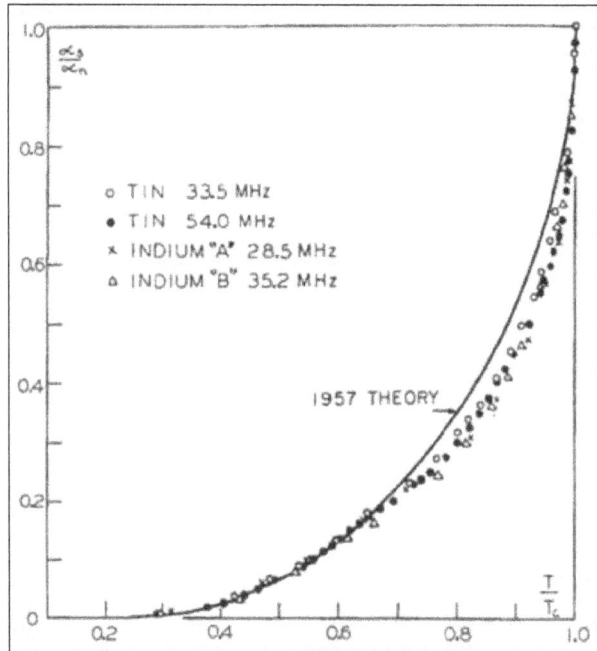

Comparison of ultrasonic attenuation observed by Morse and Bohm with theory.

Interference effects manifest themselves in a similar manner in the interaction of electromagnetic radiation with the superconductor. Near T_c the absorption is dominated by quasi-particle scattering matrix elements. Near $T=0$ the number of quasiparticle excitations goes to zero and the matrix elements that contribute are those in which quasiparticle pairs are created from ψ_0. For absorption these latter occur only when $\hbar\omega > 2\Delta$. For the linear response of the superconductor to a static magnetic field, the interference occurs in such a manner that the paramagnetic contribution goes to zero leaving the diamagnetic part which gives the Meissner effect.

It is now believed that the finite many-nucleon system that is the atomic nucleus enters a correlated state analogous to that of a superconductor. Similar considerations have been applied to many-fermion systems as diverse as neutron stars, liquid He^3 and to elementary fermions. In more recent years the importance of BCS pairing was demonstrated in ultra-cold gases. It was proposed that BCS-type physics can be responsible for some of the observed quantum Hall states. In addition the idea of spontaneously broken symmetry of a degenerate vacuum has been applied widely in elementary particle theory, in particular, in the theory of electroweak interactions. The pairing which the electron-phonon interaction has produced between electrons in metals may be produced by the van der Waals interaction between atoms in He^3, the nuclear interaction in nuclei and neutron stars, and the fundamental interactions in elementary fermions. Whatever the success of these attempts, for the theoretician the possible existence of this correlated paired state must in the future be considered for any many-fermion system where there is some kind of effective attraction between fermions for transitions near Fermi surface.

Cooper Pairs

Cooper pair is a pair of electrons (or other fermions) bound together at low temperatures in a certain manner first described in 1956 by American physicist Cooper. Cooper showed that an arbitrarily small attraction between electrons in a metal can cause a paired state of electrons to have a lower energy than the Fermi energy, which implies that the pair is bound. In conventional super-conductors, this attraction is due to the electro-phonon interaction. The Cooper pair state is responsible for superconductivity, as described in the BCS theory developed by John Bardeen, Leon Cooper, and John Schrieffer for which they shared the 1972 Nobel Prize.

Although Cooper pairing is a quantum effect, the reason for the pairing can be seen from a simplified classical explanation. An electron in a metal normally behaves as a free particle. The electron is repelled from other electrons due to their negative charge, but it also attracts the positive ions-that make up the rigid lattice of the metal. This attraction distorts the ion lattice, moving the ions slightly toward the electron, increasing the positive charge density of the lattice in the vicinity. This positive charge can attract other electrons. At long distances, this attraction between electrons due to the displaced ions can overcome the electrons' repulsion due to their negative charge, and cause them to pair up. The rigorous quantum mechanical explanation shows that the effect is due to electron-phonon interactions, with the phonon being the collective motion of the positively-charged lattice.

The energy of the pairing interaction is quite weak, of the order of 10^{-3} eV, and thermal energy can easily break the pairs. So only at low temperatures, in metal and other substrates, are a significant number of the electrons in Cooper pairs.

The electrons in a pair are not necessarily close together; because the interaction is long range, paired electrons may still be many hundreds of nanometers apart. This distance is usually greater than the average inter-electron distance; so many Cooper pairs can occupy the same space. Electrons have spin$-\frac{1}{2}$, so they are fermions, but the total spin of a Cooper pair is integer (0 or 1) so it is a composite boson. This means the wave functions are symmetric under particle interchange. Therefore unlike electrons, multiple Cooper pairs are allowed to be in the same quantum state.

The BCS theory is also applicable to other fermions systems, such as helium-3. Indeed, Cooper pairing is responsible for the superfluidity of helium-3 at low temperatures. It has also been recently demonstrated that a Cooper pair can comprise two bosons. Here, the pairing is supported by entanglement in an optical lattice.

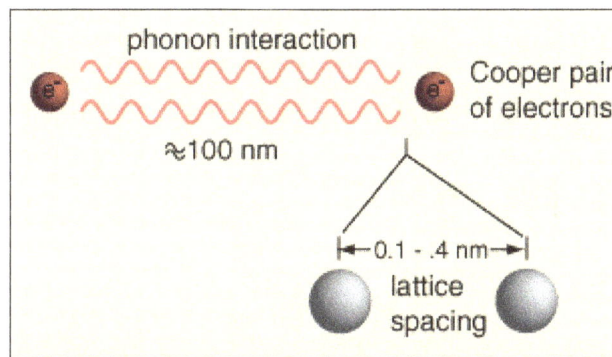

Cooper pairs.

The behavior of superconductors suggests that electron pairs are coupling over a range of hundreds of nanometers, three orders of magnitude larger than the lattice spacing called Cooper pairs, these coupled electrons can take the character of a boson and condense into the ground state.

This pair condensation is the basis for the BCS theory of superconductivity. The effective net attraction between the normally repulsive electrons produces a pair binding energy on the order of milli-electron volts, enough to keep them paired at extremely low temperatures.

The transition of a metal from the normal to the superconducting state has the nature of a condensation of the electrons into a state which leaves a band gap above them. This kind of condensation is seen with superfluid helium, but helium is made up of bosons -- multiple electrons can't collect into a single state because of the Pauli Exclusion Principle. Froehlich was first to suggest that the electrons act as pairs coupled by lattice vibrations in the material. This coupling is viewed as an exchange of phonons, phonons being the quanta of lattice vibration energy. Experimental corroboration of an interaction with the lattice was provided by the isotope effect on the superconducting transition temperature. The boson-like behavior of such electron pairs was further investigated by Cooper and they are called "Cooper pairs". The condensation of Cooper pairs is the foundation of the BCS theory of superconductivity.

If a superconductor has the form of a ring and current is once established in it by some means, the current, even several hundred amperes, flows indefinitely, even for over a year, without any change in its value. Such currents are called persistent currents.

A Model of Cooper Pair Attraction

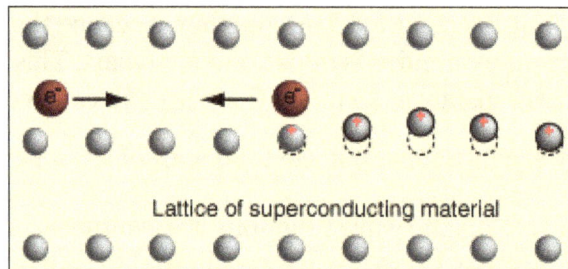

A passing electron attracts the lattice, causing a slight ripple toward its path.

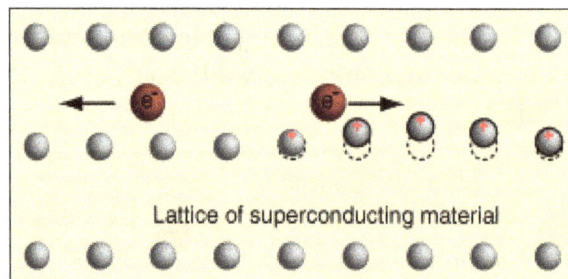

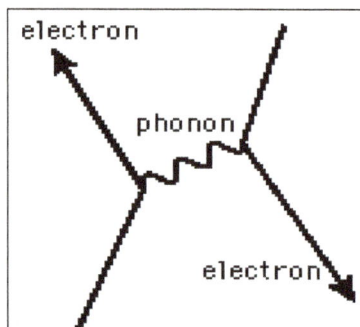

Another electron passing in the opposite direction is attracted to that displacement.

A visual model of the Cooper pair attraction has a passing electron which attracts the lattice,- causing a slight ripple toward its path. Another electron passing in the opposite direction is attracted to that displacement. This constitutes a coupling between electrons which can be depicted in a Feynman diagram. As strange as such an interaction seems,it is experimentally supported by the isotope effect and the evidence for a condensation at the critical temperature for superconductivity.

Isotope Effect and Mercury

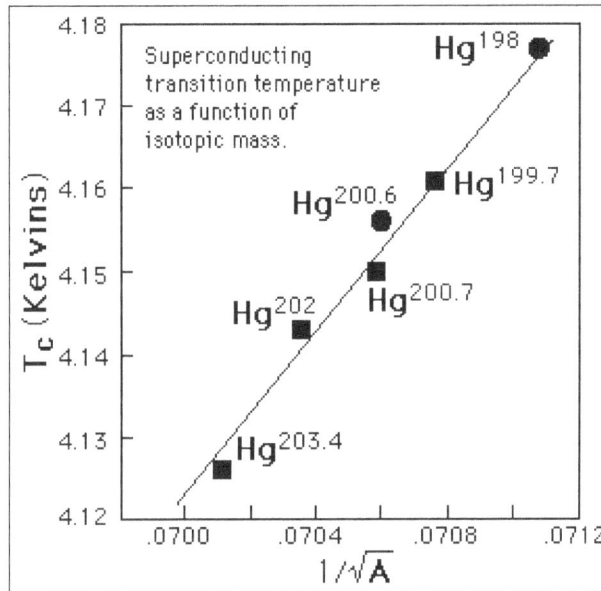

If electrical conduction in mercury were purely electronic, there should be no dependence upon the nuclear masses. This dependence of the critical temperature for superconductivity upon isotopic mass was the first direct evidence for interaction between the electrons and the lattice. This supported the BCS theory of lattice coupling of electron pairs.

It is quite remarkable that an electrical phenomenon like the transition to zero resistivity should involve a purely mechanical property of the lattice. Since a change in the critical temperature involves a change in the energy environment associated with the superconducting transition, this suggests that part of the energy is being used to move the atoms of the lattice since the energy depends upon the mass of the lattice. This indicates that lattice vibrations are a part of the superconducting process. This was an important clue in the process of developing the BCS theory because it suggested lattice coupling, and in the quantum treatment suggested that phonons were involved.

Measured Superconductor Band Gap

The measured band gap in Type I superconductors is one of the pieces of experimental evidence which supports the BCS theory. The BCS theory predicts a band gap of:

$$E_g \approx \frac{7}{2}kT_c$$

where T_c is the critical temperature for the superconductor. The energy gap is related to the coherence length for the superconductor, one of the two characteristic lengths associated with superconductivity.

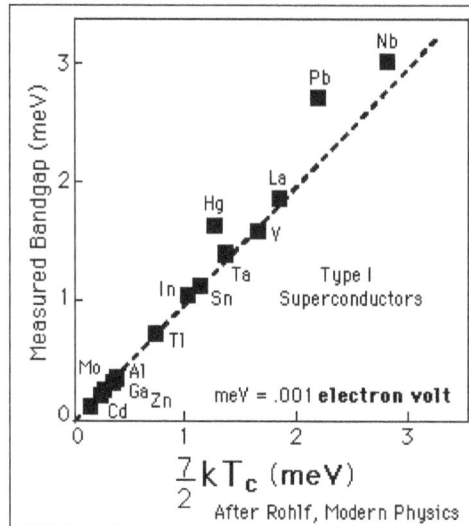

After Rohlf, Modern Physics

Energy Gap in Superconductors as a Function of Temperature

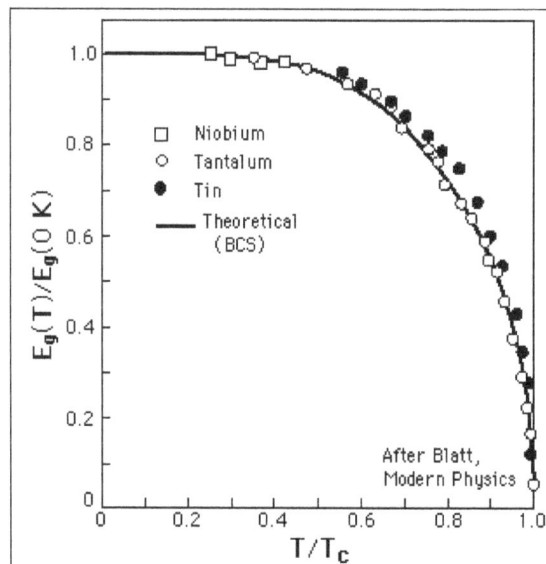

After Blatt, Modern Physics

The effective energy gap in superconductors can be measured in microwave absorption experiments. The data at left offer general confirmation of the BCS theory of superconductivity. The data is attributed to Townsend and Sutton. The reduction of the energy gap as you approach the critical temperature can be taken as an indication that the charge carriers have some sort of collective nature. That is, the charge carriers must consist of at least two things which are bound together, and the binding energy is weakening as you approach the critical temperature. Above the critical temperature, such collections do not exist, and normal resistivity prevails. This kind of evidence, along with the isotope effect which showed that the crystal lattice was involved, helped to suggest the picture of paired electrons bound together by phonon interactions with the lattice.

Bean Critical State Model in Superconductivity

The Bean critical-state model provides a phenomenological description for the hysteretic magnetization of type-II superconductors in a temporally varying external magnetic field. The magnetic field penetrates into these superconductors in the form of superconductive electron current vortices around the extremely thin filaments of normal material. Each of the magnetic vortices carries the same amount (one quantum) of magnetic flux, and so the magnetization depends on the vortex distribution. According to the Bean model, the distribution of vortices in a type-II superconductor is determined by the balance between electromagnetic driving forces and forces pinning the vortices to material inhomogeneities. Whenever the external magnetic field is changed, magnetic vortices start to enter or leave the superconductor through its boundary. If a region appears where the driving forces overcome the pinning, the system of vortices rearranges itself into another metastable state such that all vortices are pinned again and the equilibrium with the external field at the boundary is re-established. Since the unpinned vortices move rapidly, the system quickly adjusts itself to the changing external conditions, and thus a quasistationary model with instantaneous interactions is justified.

In terms of macroscopic quantities, these assumptions may be summarized by the statement that the current density never exceeds some critical value determined by the density of pinning forces and, as long as this threshold is not reached, the magnetic induction remains unchanged. There is also a caveat that if the current density is not orthogonal to the magnetic field, the equilibrium of the vortices depends on the component of current density normal to the magnetic field, and we will discuss this further below.

Together with Maxwell equations, these rules form a mathematical model of magnetization. The model, however, consists of a complicated system of equations and inequalities, and leads to a difficult free boundary problem, because the boundary between the regions of critical and subcritical currents is unknown.

To solve the problem numerically, Bossavit has recently proposed an interesting variational formulation for this free boundary problem based on the generalization of Ohm's law: the electric field was represented as the subdifferential of a non-smooth convex functional of current density. In fact, Bossavit generalized also the Bean model: he included into his formulation the material transition from the superconductive into the normal state at a point, where the effective resistivity, characterizing the energy dissipation due to the movement of vortices, reaches the value of resistivity in the normal state.

This transition has not been previously considered in the frame of Bean model, presumably because when the resistivity is high, the heat generation may become uncontrollable and cause catastrophic jumps of magnetic flux, undesirable in practical applications. Implicitly, it was assumed in the Bean model that no transition into the normal state occurs and, in particular, the effective resistivity is less than the resistivity in the normal state. Such regimes are of main interest in applications.

Model of the Critical State

Let a superconductor occupy an open bounded domain $\Omega \subset R^3$ with a Lipschitz boundary Γ and $\omega = R^3 \setminus \Omega$ be the space exterior to this domain. We start with Maxwell equations with the displacement current omitted, because the time scale for electromagnetic wave propagation is short compared to the magnetization time scale:

$$\frac{\partial \boldsymbol{B}}{\partial t} + \operatorname{curl} \boldsymbol{E} = \boldsymbol{0},$$

$$\boldsymbol{J} = \operatorname{curl} \boldsymbol{H}.$$

In the exterior ω, the constitutive equation reads $\boldsymbol{B} = \mu_0 \boldsymbol{H}$, where μ_0 is the permeability of vacuum, and the current density is given:

$$\operatorname{curl} \boldsymbol{H} = \boldsymbol{J}_e \text{ in } \omega.$$

Here $\boldsymbol{J}_e(x,t)$ is the density of external currents. It is supposed that $\operatorname{div} \boldsymbol{J}_e = 0$ and $\operatorname{supp} \boldsymbol{J}_e$ is a bounded subset of ω.

For the superconducting medium a nonlinear dependence $\boldsymbol{B} = \mu(|\boldsymbol{H}|)\boldsymbol{H}$ is supposed to be known. In the presence of electrical current flowing through the superconductor, vortices respond to the action of a Lorentz force which we average into a body force with density:

$$\boldsymbol{F}_L = \boldsymbol{J} \wedge \boldsymbol{B}.$$

Whenever the vortices become unpinned and move, they move in the direction of this force, and so their velocity v is parallel to $\boldsymbol{F}_L$. The movement of vortices induces the electric field:

$$\boldsymbol{E} = \boldsymbol{B} \wedge \boldsymbol{v},$$

which is thus parallel to $\boldsymbol{B} \wedge (\boldsymbol{J} \wedge \boldsymbol{B})$. If $\boldsymbol{B}$ is perpendicular to $\boldsymbol{J}$, as is always the case for two-dimensional problems and also for some three-dimensional, e.g., those with axial symmetry, the vectors of current density and electric field are co-linear and,

$$\boldsymbol{E} = \rho \boldsymbol{J} \text{ in } \Omega,$$

where,

$$\rho(x,t) \geq 0$$

is an unknown nonnegative function. Only the case $B \perp J$ will be considered here.

Equation $\boldsymbol{E} = \rho \boldsymbol{J}$ in Ω, may be regarded as Ohm's law with an effective resistivity ρ. However, since the resistivity is an auxiliary unknown, this relation, for a given current density, fixes only the possible direction of the electric field but not its magnitude. The next two conditions are postulates of the critical state theory:

1. The current density cannot exceed some critical value, J_c. In the Bean model of the critical state, J_c is a constant determined by the properties of the superconductive material. However, Kim et al. found that generally the critical current density depends on the magnetic field and various relations of the type $J_c = J_c(|\boldsymbol{H}|)$ have been proposed. The constraint on the current density may be written as:

$$\operatorname{curl} \boldsymbol{H}| \leq J_c(|\boldsymbol{H}|) \text{ in } \Omega.$$

We assume that,

$$\exists m, M : 0 < m < J_c(r) < M, \forall r \geq 0.$$

2. Magnetic vortices do not move in the regions where the current density is less than critical. The current in these regions is purely superconductive and the electric field must be zero. Mathematically, this can be formulated as:

$$|\operatorname{curl} \boldsymbol{H}| < J_c(|\boldsymbol{H}|) \Rightarrow \rho = 0.$$

In addition, an initial distribution of magnetic induction,

$$\boldsymbol{B}\big|_{t=0} = \boldsymbol{B}_0(x),$$

should be specified satisfying the condition div $\mathbf{B}_0 = 0$.

On the boundary dividing the two media, the tangential component of electric field E is continuous,

$$[\boldsymbol{E}_\tau] = 0 \text{ on } \Gamma,$$

where [.] denotes the jump across the boundary. We neglect the surface current, and so the tangential component of magnetic field $\boldsymbol{H}$ on this boundary is also assumed to be continuous:

$$[\boldsymbol{H}_\tau] = 0 \text{ on } \Gamma.$$

We also suppose that $|\boldsymbol{H}| \to 0$ as $|x| \to \infty$.

Quasivariational Inequality

Let us define a Hilbert space of vector functions:

$$V = \left\{ \varphi \in L^2(R^3; R^3) \middle| \operatorname{curl} \varphi \in L^2(R^3; R^3), \atop \scriptstyle \operatorname{curl} \varphi|_{\omega=0} \right\}$$

with the norm $\|\varphi\|V = \|\varphi\|L^2 + \|curl\varphi\|_{L^2}$. Note that for any $\varphi \in V$, the boundary values φ_τ on both sides of Γ are defined $H^{-1/2}(\Gamma; R^3)$ and $[\varphi\tau]|\Gamma = 0$.

Multiplying $\dfrac{\partial B}{\partial t} + \operatorname{curl} E = 0$, by $\varphi \in V$ and integrating over ω, we obtain,

$$0 = \int_\omega \left(\frac{\partial \boldsymbol{B}}{\partial t} + \operatorname{curl} \boldsymbol{E} \right) \cdot \varphi = \int_\omega \frac{\partial \boldsymbol{B}}{\partial t} \cdot \varphi + \int_\omega [\boldsymbol{E} \cdot \operatorname{curl} \varphi - \operatorname{div}(\boldsymbol{E} \wedge \varphi)]$$

$$= \int_\omega \frac{\partial \boldsymbol{B}}{\partial t} \cdot \varphi + \oint_{\Gamma+} (\boldsymbol{E} \wedge \varphi) \cdot n,$$

since curl $\varphi = 0$ in ω. Here the normal $\boldsymbol{n}$ is directed towards the domain ω. Similarly, in Ω $\dfrac{\partial \boldsymbol{B}}{\partial t} + \operatorname{curl} \boldsymbol{E} = 0$, $J = \operatorname{curl} H$. and $E = \rho J$ in Ω, yield,

$$0 = \int_{\Omega} \frac{\partial \boldsymbol{B}}{\partial t} \cdot \varphi + \int_{\Omega} \rho \operatorname{curl} H \cdot \operatorname{curl} \varphi - \oint_{\Gamma_-} (E \wedge \varphi) \cdot \boldsymbol{n}.$$

Adding the two last equations and taking into account that the tangential components of E and φ are continuous on Γ, we obtain the variational relation,

$$\int_{R^3} \frac{\partial \boldsymbol{B}}{\partial t} \cdot \varphi + \int_{\Omega} \rho \operatorname{curl} H \cdot \operatorname{curl} \varphi = 0, \quad \forall \varphi \in V.$$

The magnetic field is now determined by the boundary condition at infinity. Since $\boldsymbol{B}$ is a known function of H, the model contains only two unknowns: the magnetic field H and effective resistivity ρ. As shown below, we will be able to regard the latter function a Lagrange multiplier related to the inequality constraint $\operatorname{curl} H \leq J_c(|H|)$ in Ω.

Let us define $J_e = 0$ in Ω and consider first an auxiliary problem in R^3 :

$$\operatorname{curl} \tilde{H} = J_e,$$
$$\operatorname{div} \tilde{H} = 0,$$
$$|\tilde{H}| \to 0 \text{ as } |x| \to \infty.$$

We assume that for all t, J_e is a distribution with the compact support $\operatorname{supp} J_e \subset \omega$ and $\operatorname{div} J_e = 0$ in the space of distributions $\mathcal{D}'(R^3)$. Therefore,

$$\operatorname{curl} \tilde{H} = J_e,$$
$$\operatorname{div} \tilde{H} = 0,$$
$$|\tilde{H}| \to 0 \text{ as } |x| \to \infty.$$

has a unique solution which may be represented by means of convolution of two distributions,

$$\tilde{H} = \operatorname{curl}(\mathcal{G} * J_e),$$

where $\mathcal{G} = 1/(4\pi|x|)$ is the Green function of Laplace equation. For $J_e \in H^1(0,T;H^{-1}(R^3;R^3))$ this solution belongs to $H^1(0,T;L^2(R^3;R^3))$ and is an infinitely smooth function in $R^3 \setminus \operatorname{supp} J_e$. Hence, introducing a new variable h = H - $\tilde{H}$ and using the above equations, we obtain:

$$\operatorname{curl} h = \mathbf{0} \text{ in } \omega,$$

$$|\operatorname{curl} h| \leq J_c(|h + \tilde{H}|) \text{ in } \Omega$$

$$[h_\tau] = 0 \quad \text{on } \Gamma.$$

The only difference in the two-dimensional case is that the Green function $\mathcal{G} = -1/(2\pi)\ln(|x|)$. Below we use the following notations:

$$Q = \Omega \times (0,T), \quad \mathcal{H}L^\infty(Q), \quad \mathcal{V} = L^2(0,T;V)$$

and $\mathcal{L} = L^2(R^3 \times (0,T);R^3)$. By $\mathcal{X}'$ we denote the space dual to $\mathcal{X}$, $(\cdot,\cdot)$ means the natural pairing of the elements of $\mathcal{X}$ and $\mathcal{X}'$, and we make no distinction between Hilbert space L and its dual. We also introduce the partial ordering on the space $\mathcal{H}: \varphi \geq \psi$ if this inequality holds almost everywhere. This relation induces a partial ordering on the dual space, so that $\chi \in \mathcal{H}', \chi \geq 0$ if and only if $(\chi,\psi) \geq 0$ for all $\psi \in \mathcal{H}, \psi \geq 0$.

Let us define a family of closed convex sets of vector functions,

$$\mathcal{K}(h) = \left\{ \varphi \in \mathcal{V} \mid |\operatorname{curl}\varphi| \leq J_c\left(|h + \tilde{H}|\right) \text{ a.e. in } Q \right\},$$

and consider the quasivariational inequality,

find $h \in \mathcal{K}(h)$ such that
$$\partial\tilde{B}(h)/\partial t \in \mathcal{L},$$
$$\left(\partial\tilde{B}(h)/\partial t, \varphi - h\right) \geq 0, \forall \varphi \in \mathcal{K}(h),$$
$$\tilde{B}(h)\big|_{t=0} = B_0,$$

where $\tilde{B}(h) = B(h + \tilde{H})$.

Theorem: The function $h(x,t)$ is a solution of the quasivariational inequality if and only if there exists a functional $\rho \in \mathcal{H}'$ such that the pair $\{H,\rho\}$, where $H = h + \tilde{H}$, is a weak solution of the critical-state problem.

$$\int_{R^3} \frac{\partial B}{\partial t} \cdot \varphi + \int_\Omega \rho\,\operatorname{curl} H \cdot \operatorname{curl}\varphi = 0, \quad \forall \varphi \in V.$$

Proof: For any function $h \in \mathcal{V}$ such that $\partial\tilde{B}(h)/\partial t \in \mathcal{L}$ we define the linear functional,

$$F_h(\varphi) = \left(\partial\tilde{B}(h)/\partial t, \varphi\right)$$

and the nonlinear operator $G_h : \mathcal{A} \to \mathcal{H}$,

$$G_h(\varphi) = \frac{1}{2}\left(|\operatorname{curl}\varphi|^2 - J_c^2\left(|h + \tilde{H}|\right)\right)\Big|_\varrho,$$

where,

$$\mathcal{A} = \{\varphi \in \mathcal{V} \mid |\operatorname{curl}\varphi| \leq 2M \text{ a.e. in } Q\}$$

is a closed convex set. Now the inequality,

$$\text{find } h \in \mathcal{K}(h) \text{ such that}$$
$$\partial \tilde{B}(h) / \partial t \in \mathcal{L},$$
$$\left(\partial \tilde{B}(h) / \partial t, \varphi - h \right) \geq 0, \forall \varphi \in \mathcal{K}(h),$$
$$\tilde{B}(h) \big|_{t=0} = B_0,$$

can be formally written as an optimization problem,

$$h \in \arg \min \; Fh(\varphi).$$
$$G_h(\varphi) \leq 0$$
$$\varphi \in \mathcal{A}$$

This representation allows us to introduce a Lagrange multiplier into the quasivariational inequality. To do this, let us fix h in F_h and G_h. The continuous functional F_h is linear, and the mapping G_h is convex in the sense of the partial ordering on $\mathcal{H}$ defined above. The cone $\mathcal{C}$ of nonnegative elements in $\mathcal{H}$ has a non-empty interior. Since $J_c > m > 0$ for any h, the constraint qualification hypothesis $\exists \varphi_0 \in \mathcal{A} : -G_h(\varphi_0) \in \text{int} \, \mathcal{C}$ is satisfied with $\varphi_0 \equiv 0$.

If the functional F_h is bounded from below in $\mathcal{K}(h)$, and we will check this later, the condition of optimality for,

$$h \in \arg \min \; Fh(\varphi).$$
$$G_h(\varphi) \leq 0$$
$$\varphi \in \mathcal{A}$$

may be obtained using the Lagrange multiplier technique $\left[H_\tau \right] = 0$ on Γ.

u is a point of minimum if and only if there exists a Lagrange multiplier $\rho \in \mathcal{H}', \rho \geq 0$, such that the pair $\{u, \rho\}$ is a saddle point of the Lagrangian, i.e.,

$$F_h(u) + (\rho^*, G_h(u)) \leq F_h(u) + (\rho, G_h(u)) \leq F_h(h^*) + (\rho, G_h(h^*))$$

for all $h^* \in \mathcal{A}, \rho^* \in \mathcal{H}', \rho^* \geq 0$.

In our case both functional and constraint depend on the unknown solution. However, if this solution exists, it should satisfy the optimality condition above. Therefore, substituting $u = h$, we obtain the condition of optimality for the implicit optimization problem:

$$h \in \arg \min \; Fh(\varphi).$$
$$G_h(\varphi) \leq 0$$
$$\varphi \in \mathcal{A}$$

and are now able to formulate the following saddle-point condition for the quasivariational inequality: is a solution of,

$$\text{find } h \in \mathcal{K}(h) \text{ such that}$$

$$\partial \tilde{B}(h) / \partial t \in \mathcal{L},$$

$$\left(\partial \tilde{B}(h) / \partial t, \varphi - h\right) \geq 0, \forall \varphi \in \mathcal{K}(h),$$

$$\tilde{B}(h)\big|_{t=0} = B_0,$$

if and only if it satisfies the initial condition, $\partial \tilde{B}(h) / \partial t \in \mathcal{L}$, and there exists a Lagrange multiplier $\rho \in \mathcal{H}', \rho \geq 0$, such that the pair $\{h, \rho\}$ is a saddle point of the Lagrangian, i.e.,

$$F_h(h) + (\rho^*, G_h(h)) \leq F_h(h) + (\rho, G_h(h)) \leq F_h(h^*) + (\rho, G_h(h^*))$$

for all $h^* \in \mathcal{A}, \rho^* \in \mathcal{H}', \rho^* \geq 0$. Note that from the first inequality in,

$$F_h(h) + (\rho^*, G_h(h)) \leq F_h(h) + (\rho, G_h(h)) \leq F_h(h^*) + (\rho, G_h(h^*))$$

follows that $G_h(h) \leq 0$ and,

$$\left(\rho, G_h(h)\right) = 0$$

This is the complementary slackness condition, which means that the Lagrange multiplier may be nonzero only where the constraint is active.

Let h be a solution of,

$$\text{find } h \in \mathcal{K}(h) \text{ such that}$$

$$\partial \tilde{B}(h) / \partial t \in \mathcal{L},$$

$$\left(\partial \tilde{B}(h) / \partial t, \varphi - h\right) \geq 0, \forall \varphi \in \mathcal{K}(h),$$

$$\tilde{B}(h)\big|_{t=0} = B_0,$$

Then the functional   is bounded from below in $\mathcal{K}(h)$, and so there exists a saddle point $\{h, \rho\} \in \mathcal{V} \times \mathcal{H}'$. Let us set $H = h + \tilde{H}$ and show that $\{H, \rho\}$ is a weak solution to the critical-state problem. Since $h \in \mathcal{K}(h)$, $\operatorname{curl} H = J_e$ in ω $\operatorname{curl} H | \leq J_c(|H|)$ in Ω and $[H_\tau] = 0$ on Γ hold in the sense of distributions or almost everywhere. The inequality $\rho(x,t) \geq 0$ is true in the sense of the partial ordering on $\mathcal{H}'$.

Like any nonnegative functional in $\mathcal{H}', \rho$ may be represented as:

$$(\rho, \psi) = \int_Q \psi d\mu,$$

where μ is a nonnegative additive function defined on the Lebesgue-measurable subsets of Q and such that $\mu(Q) < \infty$ and $\operatorname{mes}(\tilde{Q}) = 0 \Rightarrow \mu(\tilde{Q}) = 0, \forall \tilde{Q} \subset Q$. Define,

$$Q^- = \operatorname{supp} G_h(h) = \{x, t\} \in Q \big| |\operatorname{curl} H| < J_c(|H|) \, a.e.\},$$

the complementarity slackness condition $(\rho, G_h(h)) = 0$ yields,

$$\int_{Q^-} G_h(h) d\mu = 0.$$

Since $G_h(h) < 0$ a.e. in Q^- we conclude that $\mu(Q^-) = 0$, and so $|\mathrm{curl}\, H| < J_c(|H|) \Rightarrow \rho = 0$ holds in the following weak sense:

$$\forall \psi \in \mathcal{H} \,\mathrm{supp}\, \psi \subset Q^- \Rightarrow (\rho, \psi) = 0.$$

The second inequality in $F_h(u) + (\rho^*, G_h(u)) \le F_h(u) + (\rho, G_h(u)) \le F_h(h^*) + (\rho, G_h(h^*))$ means that the functional $F_h(h^*) + (\rho, G_h(h^*))$ attains a minimum on $\mathcal{A}$ at the point $h^* = h$. We now define,

$$\mathcal{W} = \{\varphi \in \mathcal{V} \mid |\mathrm{curl}\, \varphi| \in \mathcal{H}.$$

Since $|\mathrm{curl}\, h| < M$ a.e. in Q, for any function $\varphi \in \mathcal{W}$ there exists $\varepsilon_0 > 0$ such that $h + \varepsilon \varphi \in \mathcal{A}$ for all $\varepsilon \in (\varepsilon_0, \varepsilon_0)$. Therefore,

$$F_h(\varphi) + (\rho, \mathrm{curl}\, \varphi) = 0, \forall \varphi \in \mathcal{W}.$$

The functions $\mathrm{curl}\, h$ and $\mathrm{curl}\, H$ coincide in Ω, so this is equivalent to a weak form of:

$$\int_{R^3} \frac{\partial B}{\partial t} \cdot \varphi + \int_\Omega \rho\, \mathrm{curl}\, H \cdot \mathrm{curl}\, \varphi = 0, \quad \forall \varphi \in V$$

$$(\partial B(H)/\partial t, \varphi) + (\rho, \mathrm{curl}\, H \cdot \mathrm{curl}\, \varphi) = 0, \quad \forall \varphi \in \mathcal{W}.$$

This proves that $\{H, \rho\}$ is a weak solution to problem $\mathrm{curl}\, H = J_e$ in ω, $\rho(x,t) \ge 0$,

$$\int_{R^3} \frac{\partial B}{\partial t} \cdot \varphi + \int_\Omega \rho\, \mathrm{curl}\, H \cdot \mathrm{curl}\, \varphi = 0, \quad \forall \varphi \in V.$$

Now let $\{H, \rho\} \in \mathcal{L} \times \mathcal{H}'$ be a weak solution of the critical-state model (in the sense clarified in the first part of this proof) and $\partial B(H)/\partial t \in \mathcal{L}$. We need to show that $h = H - \tilde{H}$ is a solution of the quasivariational inequality

     find $h \in \mathcal{K}(h)$ such that
$$\partial \tilde{B}(h)/\partial t \in \mathcal{L},$$
$$\left(\partial \tilde{B}(h)/\partial t, \varphi - h\right) \ge 0, \forall \varphi \in \mathcal{K}(h),$$
$$\tilde{B}(h)\big|_{t=0} = B_0,$$

Equation $[h_\tau] = 0$ on Γ, understood in the sense of distributions, and $\mathrm{curl}\, h = 0$ in ω, $|\mathrm{curl}\, h| \le J_c(|h + \tilde{H}|)$ in Ω satisfied almost everywhere, yield $\mathrm{curl}\, h \in L^2(R^3; R^3)$ for almost every t and $h \in \mathcal{K}(h)$. It follows from the weak version of $|\mathrm{curl}\, H| < J_c(|H|) \Rightarrow \rho = 0$ that $(\rho, G_h(h)) = 0,$.

Hence,

$$\left(\rho, |\mathrm{curl}\, h|^2\right) = \left(\rho, J_c^2\left(h + \tilde{H}\right)\right).$$

Let $\varphi \in \mathcal{K}(h)$, using equations $\left(\partial B(H)/\partial t, \varphi\right) + \left(\rho, \mathrm{curl}\, H \cdot \mathrm{curl}\, \varphi\right) = 0$, $\forall \varphi \in W$, which is the weak form of $\int_{R^3} \dfrac{\partial B}{\partial t} \cdot \varphi + \int_{\Omega} \rho\, \mathrm{curl}\, H \cdot \mathrm{curl}\, \varphi = 0$, $\forall \varphi \in V$ and $\left(\rho, |\mathrm{curl}\, h|^2\right) = \left(\rho, J_c^2\left(h + \tilde{H}\right)\right)$. We obtain

$$\left(\partial \tilde{B}(h)/\partial t, \varphi - h\right) = -(\rho, \mathrm{curl}\, h \cdot \mathrm{curl}\{\varphi - h\})$$
$$= \left(\rho, J_c^2(h + \tilde{H})\right) - \mathrm{curl}\, h \cdot \mathrm{curl}\, \varphi\Big) \geq 0,$$

since $\varphi, h \in \mathcal{K}(h)$. Thus the theorem is proved.

Variational Inequality

If, as is assumed in the Bean model, the critical current density J_c does not depend on the magnetic field, we have $\mathcal{K}(h) \equiv \mathcal{K}$ and the inequality:

$$\text{find } h \in \mathcal{K}(h) \text{ such that}$$
$$\partial \tilde{B}(h)/\partial t \in \mathcal{L},$$
$$\left(\partial \tilde{B}(h)/\partial t, \varphi - h\right) \geq 0, \forall \varphi \in \mathcal{K}(h),$$
$$\tilde{B}(h)\big|_{t=0} = B_0,$$

becomes a variational one.

Also, another simplifying assumption, $\mu(|H|) \approx \mu_0$, is justified for strong fields. Under these two assumptions, the variational inequality may be written as follows:

$$h(.,t) \in K : \left(\partial h/\partial t - f, \varphi - h\right) \geq 0, \forall \varphi \in K,$$
$$h\big|_{t=0} = h_0.$$

Here $f = -\partial \tilde{H}/\partial t \in \mathcal{L}$, $h_0 = B_0/\mu_0 - \tilde{H}\big|_{t=0}$,

and

$$K = \{\varphi \in V \mid |\mathrm{curl}\, \varphi| \leq J_c \text{ a.e. in } \Omega\}.$$

Theorem: Let $h_0 \in K$. Then the variational inequality,

$$h(.,t) \in K : \left(\partial h/\partial t - f, \varphi - h\right) \geq 0, \forall \varphi \in K,$$
$$h\big|_{t=0} = h_0$$

has a unique solution $h \in C([0,T]); L^2(R^3; R^3)$ such that $h(.,t) \in K$ for almost all t and $\partial h / \partial t \in \mathcal{L}$. Also, $\operatorname{div} h = 0$ $a.e$ and $h(.,t) \in H^1(R^3; R^3)$ for almost all t.

Proof: The set K is a closed convex subset of the Hilbert space $L^2(R^3; R^3)$ and we can rewrite the variational inequality as a Cauchy problem:

$$dh / dt + \partial I_K(h) \ni f,$$
$$h\big|_{t=0} = h_0,$$

where ∂I_K is the subdifferential of the indicator function,

$$I_K(\psi) = \begin{cases} 0 & \text{if } \psi \in K, \\ \infty & \text{otherwise.} \end{cases}$$

Since I_K is a lower-semicontinuous convex function defined on $L^2(R^3; R^3)$ the first part of the theorem follows now immediately from the known results on differential equations with maximal monotone operators.

Let us check that $\operatorname{div} h = 0$. For any $\chi \in \partial I_K(h(.,t))$,

$$(\chi, \varphi - h) \le 0, \forall \varphi \in K.$$

By definition, $(\operatorname{div} \chi, \psi) = (\chi, -\operatorname{grad} \psi), \forall \psi \in \mathcal{D}(R^3)$. Since $\operatorname{curl} \operatorname{grad} \psi$ is zero, function $\varphi = -\operatorname{grad} \psi$ belongs to K. Using $(\chi, \varphi - h) \le 0, \forall \varphi \in K$. we obtain,

$$(\operatorname{div} \chi, \psi) \le (\chi, h), \forall \psi \in \mathcal{D}(R^3).$$

This means that $\operatorname{div} \chi = 0$ a.e. for any $\chi \in \partial I_K(h)$. Since $\tilde{H}$ is a solenoidal vector field, also $\operatorname{div} f = 0$ and $dh / dt + \partial I_K(h) \ni f$, yields $\operatorname{div} h \equiv \operatorname{div} h_0$. We have assumed that $\operatorname{div} B_0 = 0$, hence

$$h\big|_{t=0} = h_0,$$

$\operatorname{div} h_0 = 0$. We proved that $\operatorname{div} h = 0$ and, since h and $\operatorname{curl} h$ belong to $L^2(R^3; R^3)$ for almost all t, we have also $h(.,t) \in H^1(R^3; R^3)$, which completes the proof.

Corollary: The inequality $h(.,t) \in K: (\partial h / \partial t - f, \varphi - h) \ge 0, \forall \varphi \in K,$ is equivalent to the variational inequality:
$$h\big|_{t=0} = h_0.$$

$$h(.,t) \in K_0: (\partial h / \partial t - f, \varphi - h) \ge 0, \forall \varphi \in K_0,$$
$$h\big|_{t=0} = h_0,$$

with the set,

$$K_0 = \left\{ \varphi \in H^1(R^3; R^3) \middle| \begin{array}{l} \|\operatorname{curl} \varphi\| \le J_c \text{ a.e. in } \Omega, \\ \operatorname{curl} \varphi = 0 \text{ a.e. in } \omega, \\ \operatorname{div} \varphi = 0 \text{ a.e. in } R^3 \end{array} \right\}.$$

Proof: We have proved that h, the unique solution of,

$$h(.,t) \in K : \quad (\partial h / \partial t - f, \varphi - h) \geq 0, \forall \varphi \in K,$$
$$h\big|_{t=0} = h_0.$$

belongs to K_0 for almost all t. Clearly, h is a solution of,

$$h(.,t) \in K_0 : \quad (\partial h / \partial t - f, \varphi - h) \geq 0, \forall \varphi \in K_0,$$
$$h\big|_{t=0} = h_0,$$

as well, because $K_0 \subset K$. Since $h(.,t) \in K_0 : \quad (\partial h / \partial t - f, \varphi - h) \geq 0, \forall \varphi \in K_0,$

$$h\big|_{t=0} = h_0,$$

also has only one solution, the two inequalities are equivalent.

The numerical solution of variational inequalities,

$$h(.,t) \in K : \quad (\partial h / \partial t - f, \varphi - h) \geq 0, \forall \varphi \in K,$$
$$h\big|_{t=0} = h_0.$$

or

$$h(.,t) \in K_0 : \quad (\partial h / \partial t - f, \varphi - h) \geq 0, \forall \varphi \in K_0,$$
$$h\big|_{t=0} = h_0,$$

can be obtained by means of discretization and solution of convex programming problems, arising at each time layer. Solution of the quasivariational inequality would need an additional level of iterations. However, the realization of this procedure is difficult because the unknown magnetic field must be calculated in the whole space. To avoid this difficulty, we now derive a variational formulation in terms of the current density.

Obstacle Problem

Let us define a closed convex set,

$$K_1 = \left\{ \psi \in L^2\left(R^3; R^3\right) \middle| \begin{array}{l} \|\psi\| \leq J_c \text{ a.e. in } \Omega, \\ \psi = 0 \text{ a.e. in } \omega, \\ \text{div} \psi = 0 \text{ a.e. in } R^3 \end{array} \right\}.$$

For any $\psi \in K_1$, the function $\mathcal{R}\psi = \text{curl}(\mathcal{G} * \psi)$, where $\mathcal{G}$ is the Green function, is the only solution of the problem,

$$\text{curl}\, \varphi = \psi,$$
$$\text{div}\, \varphi = 0,$$
$$|\varphi| \to 0 \text{ as } |x| \to \infty.$$

Therefore, $\mathcal{R}\psi \in K_0$. On the other hand, $\operatorname{curl}\varphi \in K_1$ for any $\varphi \in K_0$. The linear operator $\mathcal{R}$ is inverse to the operator curl and establishes a one-to-one correspondence between the sets K_0 and K_1. Since $\tilde{H} = \mathcal{R}J_e$, we can now rewrite,

$$h(.,t) \in K_0: \ (\partial h / \partial t - f, \varphi - h) \geq 0, \forall \varphi \in K_0,$$

as,

$$h\big|_{t=0} = h_0,$$

$$J(.,t) \in K_1: \left(\mathcal{R}\{\partial J / \partial t + \partial J_e / \partial t\}, \mathcal{R}\psi - \mathcal{R}J \right), \forall \psi \in K_1,$$

$$\mathcal{R}J\big|_{t=0} = h_0,$$

or, equivalently,

$$J(.,t) \in K_1: \left(\mathcal{R}^* \mathcal{R}\{\partial J / \partial t + \partial J_e / \partial t\}, \psi - J \right), \forall \psi \in K_1,$$

$$J\big|_{t=0} = \operatorname{curl} h_0,$$

where $J = \operatorname{curl} h$ is the current density and $\mathcal{R}^*$ is ajoint to $\mathcal{R}$. Let $\mathbf{\Phi} \in H^{-1}\left(R^3; R^3 \right)$ be a distribution with the compact support, $\mathbf{\Psi} \in \mathcal{D}\left(R^3, R^3 \right)$, and $\operatorname{div} \mathbf{\Phi} = \operatorname{div} \mathbf{\Psi} = 0$. Making use of Green theorem, we obtain,

$$\left(\mathcal{R}^* \mathcal{R}\mathbf{\Phi}, \mathbf{\Psi} \right) = \left(\mathcal{R}\mathbf{\Phi}, \mathcal{R}\mathbf{\Psi} \right) = \int_{R^3} \operatorname{curl}(\mathcal{G} * \mathbf{\Phi}) \cdot \operatorname{curl}(\mathcal{G} * \mathbf{\Psi}) \ =$$

$$\int_{R^3} \mathcal{G} * \mathbf{\Phi} \cdot \operatorname{curl} \wedge \operatorname{curl}(\mathcal{G} * \mathbf{\Psi}) = \int_{R^3} \mathcal{G} * \mathbf{\Phi} \ \cdot \left[\operatorname{grad} \operatorname{div}(\mathcal{G} * \mathbf{\Psi}) - 4(\mathcal{G} * \mathbf{\Psi}) \right] =$$

$$= \int_{R^3} \mathcal{G} * \mathbf{\Phi} \cdot \left[\mathcal{G} * (\operatorname{grad} \operatorname{div} \mathbf{\Psi}) - (\Delta\mathcal{G}) * \mathbf{\Psi} \right] = \int_{R^3} \mathcal{G} * \mathbf{\Phi} \cdot \mathbf{\Psi},$$

since $\operatorname{div} \mathbf{\Psi} = 0$ and $-\Delta\mathcal{G}$ is the delta function. Thus $\mathcal{R}^* \mathcal{R}\mathbf{\Phi} = \mathcal{G} * \mathbf{\Phi}$, which is the magnetic vector potential of current $\mathbf{\Phi}$. Using this formula, we arrive at the variational inequality,

$$J \in K_1: \left(\mathcal{G} * \{\partial(J + J_e) / \partial t\}, \varphi - J \right) \geq 0, \ \forall \varphi \in K_1,$$

$$J\big|_{t=0} = \operatorname{curl} h_0.$$

A variational inequality with the similar pseudodifferential operator arises in elasticity (the stamp problem).

In two-dimensional problems,

$$\mathbf{H} \ = \ \left(H_1(x_1, \, x_2, \, t), \, H_2(x_1, \, x_2, \, t), \, 0 \right)$$

and so $J = \left(0, \, 0, \, J(x_1, \, x_2, \, t) \right)$. The divergence of current density is automatically zero, the variational inequality becomes scalar and can be written as,

$$J(.,t) \in K_2: \left(\mathcal{G} * \{\partial(J + J_e) / \partial t\}, \varphi - J \right) \geq 0, \ \forall \varphi \in K_2,$$

$$J\big|_{t=0} = J_0,$$

where,

$$K_2 = \left\{ \varphi \in L^2(\Omega) \mid \ |\varphi| \le J_c \ a.e. \right\}.$$

A similar scalar variational inequality arises in three-dimensional problems with axial symmetry. The formulations obtained can serve a basis for effective numerical method for solving the critical-state problems.

Superconductor: A Comprehensive Study

Any material which conducts electric current without any resistance is referred to as a superconductor. There are different materials that exhibit superconductivity such as mercury, lead, niobium-titanium, magnesium diboride, etc. These are further classified as type 1 and 2 superconductors. The topics elaborated in this chapter will help in gaining a better perspective about superconductors.

A superconductor is an element or metallic alloy which, when cooled below a certain threshold temperature, the material dramatically loses all electrical resistance. In principle, superconductors can allow electrical current to flow without any energy loss (although, in practice, an ideal superconductor is very hard to produce). This type of current is called a supercurrent.

The threshold temperature below which a material transitions into a superconductor state is designated as T_c, which stands for critical temperature. Not all materials turn into superconductors, and the materials that do each have their own value of T_c.

Superconductor Metals

Some metals when they are cooled below their critical temperature exhibits the zero resistivity or infinite conductivity. These metals are called superconductor metals. Some metals showing superconductivity and their critical temperatures/transition temperature are listed in table below:

SL	Superconductor	Chemical Symbol	Critical/Transition Temperature T_c(K)	Critical Magnetic Field B_c(T)
1	Rhodium	Rh	0	0.0000049
2	Tungsten	W	0.015	0.00012
3	Beryllium	Be	0.026	
4	Iridium	Ir	0.1	0.0016
5	Lutetium	Lu	0.1	
6	Hafnium	Hf	0.1	
7	Ruthenium	Ru	0.5	0.005
8	Osmium	Os	0.7	0.007
9	Molybdenum	Mo	0.92	0.0096
10	Zirconium	Zr	0.546	0.0141
11	Cadmium	Cd	0.56	0.0028
12	Uranium	U	0.2	
13	Titanium	Ti	0.39	0.0056
14	Zinc	Zn	0.85	0.0054
15	Gallium	Ga	1.083	0.0058
16	Gadolinium	Gd	1.1	
17	Aluminium	Al	1.2	0.010

18	Protactinium	Pa	1.4	
19	Thorium	Th	1.4	0.013
20	Rhenium	Re	1.4	0.030
21	Thallium	Tl	2.39	0.018
22	Indium	In	3.408	0.028
23	Tin	Sn	3.722	0.030
24	Mercury	Hg	4.153	0.040
25	Tantalum	Ta	4.47	0.083
26	Vanadium	V	5.38	0.031
27	Lanthanum	La	6.0	0.11
28	Lead	Pb	7.193	0.080
29	Technetium	T_c	7.77	0.040
30	Niobium	Nb	9.46	0.820

Types of Superconductors

Depending upon their behavior in an external magnetic field, superconductors are divided into two types:

Type I Superconductors

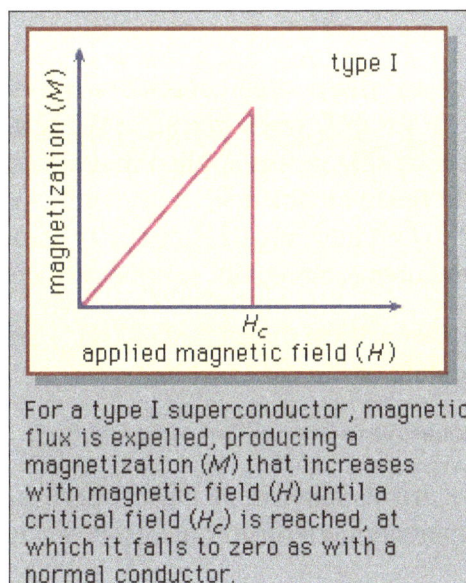

For a type I superconductor, magnetic flux is expelled, producing a magnetization (M) that increases with magnetic field (H) until a critical field (H_c) is reached, at which it falls to zero as with a normal conductor.

- Type I superconductors are those superconductors which lose their superconductivity very easily or abruptly when placed in the external magnetic field. As you can see from the graph of intensity of magnetization (M) versus applied magnetic field (H), when the Type I superconductor is placed in the magnetic field, it suddenly or easily loses its superconductivity at critical magnetic field (H_c) (point A).

- After H_c, the Type I superconductor will become conductor.

- Type I superconductors are also known as soft superconductors because of this reason that is they lose their superconductivity easily.

- Type I superconductors perfectly obey Meissner effect.

- Example of Type I superconductors: Aluminum ($H_c = 0.0105$ Tesla), Zinc ($H_c = 0.0054$).

Type II Superconductors

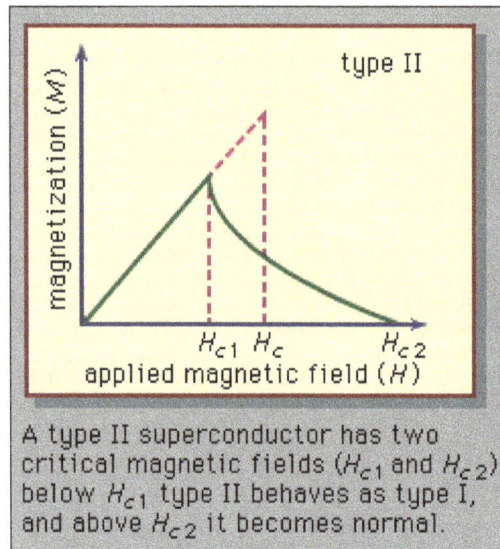

A type II superconductor has two critical magnetic fields (H_{c1} and H_{c2}); below H_{c1} type II behaves as type I, and above H_{c2} it becomes normal.

- Type II superconductors are those superconductors which lose their superconductivity gradually but not easily or abruptly when placed in the external magnetic field. The graph of intensity of magnetization (M) versus applied magnetic field (H), when the Type II superconductor is placed in the magnetic field, it gradually loses its superconductivity. Type II superconductors start to lose their superconductivity at lower critical magnetic field (H_{c1}) and completely lose their superconductivity at upper critical magnetic field (H_{c2}).

- The state between the lower critical magnetic field (H_{c1}) and upper critical magnetic field (H_{c2}) is known as vortex state or intermediate state.

- After H_{c2}, the Type II superconductor will become conductor.

- Type II superconductors are also known as hard superconductors because of this reason that is they lose their superconductivity gradually but not easily.

- Type II superconductors obey Meissner effect but not completely.

- Example of Type II superconductors: NbN ($H_c = 8 \times 10^6$ Tesla), Babi$_3$ ($H_c = 59 \times 10^3$ Tesla).

- Application of Type II superconductors: Type II superconductors are used for strong field superconducting magnets.

Superconducting Energy Gap

The superconducting energy gap can be defined as the energy difference between the ground state of the superconductor and the energy of the lowest quasiparticle excitation. There were early hints that such a gap existed but the first experimental evidence for a gap came from the temperature dependence of the specific heat below the transition temperature T_c as measured by Corak et al. It was found that the electronic specific heat was given by $C_e \propto \gamma T_c e^{-1.5 T_c/T}$ where γ is the normal state electronic specific heat coefficient and T_c the superconducting transition temperature. The first spectroscopic measurement of the energy gap was carried out with microwaves by Biondi et al on aluminum and far infrared techniques by Glover et al on lead. At the same time, the microscopic theory of superconductivity, the BCS theory, was announced by Bardeen, Cooper and Schrieffer who predicted the value of the ratio of the superconducting gap to the transition temperature to be 1.76 which is in excellent agreement with the spectroscopic measurements. This was one of the earliest triumphs of the theory. It should be noted that this value of the gap is from the weak coupling limit of the theory and applies to materials with low T_c's such as aluminum. For conventional superconductors, tunnelling spectroscopy has been a popular tool for gap measurement but for the new high temperature superconductors, with their larger gaps, infrared and photoemission spectroscopies have played an increasingly important role.

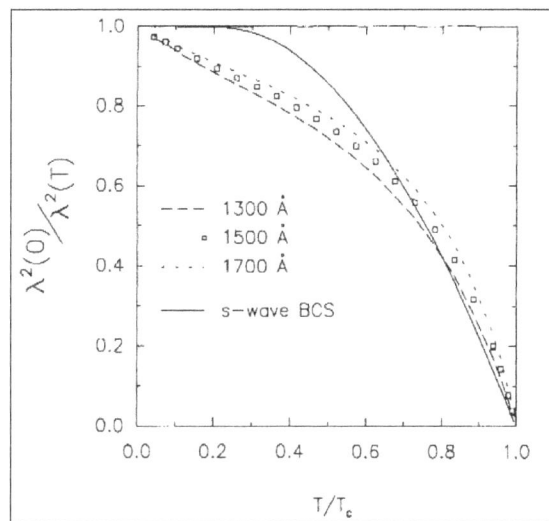

Inverse square of the superconducting penetration depth as function of temperature from Hardy et al. The linear T dependence at low temperature is evidence for a d-wave order parameter in contrast to the s-wave exponential dependence shown as the solid line.

A superconducting gap has a profound influence on the response of a superconductor to an alternating electromagnetic field. An incoming photon can be absorbed only if its energy and momentum can be transferred to the superconductor. Energy conservation demands that the photon energy $h\nu > 2\Delta$, twice the energy gap. The factor two comes from the destruction of a Cooper pair to create a pair of quasiparticles. A second requirement is the conservation of momentum. By creating a bosonic excitation, typically a phonon or a spin fluctuation, or by elastic scattering with a static defect, one of the quasiparticles can be scattered away from the Fermi surface and its momentum transferred from the superconducting condensate to other degrees of freedom. For elastic scattering to dominate, the scattering rate has to be high enough to reach the dirty limit, defined as $1/\tau$

> 2Δ where τ is the quasiparticle life time. In the dirty limit there is a sharp onset of absorption for frequencies greater than 2Δ and the optical method is a good way of determining the energy gap in the extreme case where $1/\tau \ll 2\Delta$, additional energy has to be supplied to create the bosonic excitation and the onset of absorption occurs at hν > 2Δ + $\hbar\Omega_E$ for a single Einstein boson at $\hbar\Omega E$. According to a theorem of Anderson, in the dirty limit, the superconducting gap has the same value for all points on the Fermi surface, in other words it is isotropic. The gap width in pure materials can show considerable anisotropy in momentum space as shown by Richards through infrared measurements on single crystals of Sn.

With the discovery of high temperature superconductivity by Bednorz and Müller one of the first questions raised was the nature of the superconducting gap. After an initial period of confusion, it was found that the gap in these materials had several novel properties. Early infrared spectroscopy showed that the scattering rate was very low placing the materials in the clean limit, which made the determination of the gap width difficult. Microwave measurements by the University of British Columbia group of the penetration depth variation with temperature are shown in figure. This shows conclusively that the gap was highly anisotropic in momentum space, going to zero for electrons travelling in certain directions, and yielding a gap magnitude that varied as $\left(k_x^2 - k_y^2\right)$ around the Fermi surface. This corresponds to a d-wave superconducting state with nodes as opposed to the isotropic s-wave gap of the conventional superconductors. Calculations show that there is no onset of absorption at 2Δ in the d-wave superconductor as there is in the dirty s-wave case. The final result is that while there is a notable onset of absorption the frequency of this feature (at 500 cm^{-1} in this example) is the combined energy of the gap and the boson mode that acts as the glue binding the superconducting carriers.

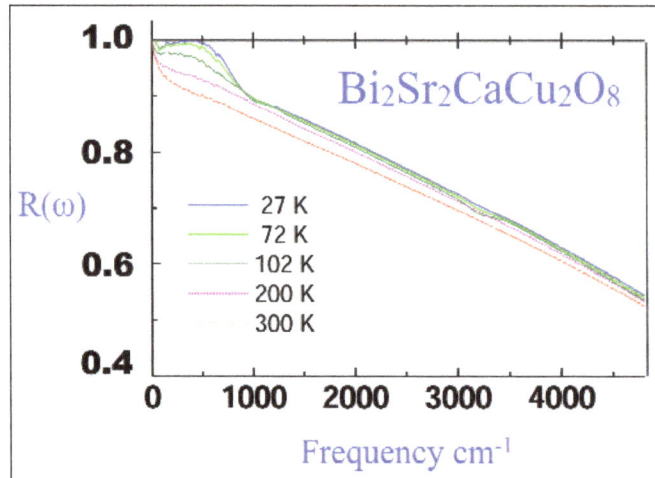

The reflectance of a high temperature superconductor with T_c = 91 K at various temperatures from Hwang et al Note the sharp onset of absorption at 27 K in the superconducting state, above a frequency of 500 cm^{-1}. In this clean limit superconductor the onset marks the energy where the incoming photon can break a Cooper pair and generate a bosonic excitation.

Another puzzle surrounding the nature of the gap of the high temperature superconductors came from several hints that the gap remained in the normal state at temperatures above the superconducting transition temperature. The first evidence from this came from NMR experiments which found a gaplike depression of the density of states at the Fermi surface below a temperature T* which was larger than T_c at lower doping levels but approached T_c near optimal doping. It was initially called the "spin gap". Subsequent experiments on the optical conductivity showed that the gap involved charge degrees of freedom and was renamed the pseudogap. A number of experiments, including specific heat, confirmed these early results. Among them is angle resolved photo emission that showed the pseudogap had the same d-wave symmetry as the superconducting gap.

Temperature Dependence of the Superconductor Energy Gap

Magnitude of the temperature dependent energy gap, the gap between the energy of the Fermi level and the next available electronic energy level in the system, for a superconductor depends strongly upon the superconductor's internal magnetic field. Because of this it is necessary to specify the internal magnetic field of a superconductor to make coherent remarks concerning its temperature dependent energy gap. For a superconductor to be able to receive an external magnetic field, there must be a vacant energy state in the superconductor to receive the energy associated with the field. For a small range of energies near that of the critical magnetic field, H_c, these energy states lie within the superconductor temperature dependent energy gap.

Balance in the loss of dissipative electron scattering and the change in entropy of the conducting phase that occur in the phase transition between the normal metal and the N superconducting state to suggest that changes in electron Gibbs free energy at T, from these sources are the basis for the temperature dependent energy gap. The critical magnetic field for a superconductor at temperature, T, $H_c(T)$ occurs when the energy of the magnetic field is equal to the magnitude of the superconductor energy gap at T. Under these conditions, the internal magnetic field of the superconductor corresponds to H_c. When the superconductor energy gap is occupied with the energy of the external magnetic field, the normal metal conducting bands that became inaccessible at the superconductor — normal conductor phase transition are once again available for conduction, and the superconductor quenches. Origins of the superconductor energy gap arising from loss of dissipative electron scattering and development of coherent electron lattice order at the superconductor phase transition lie in the laws of thermodynamics, which cannot be casually neglected. Experimental data from the literature suggests that the ratio of the superconductor energy gap to the superconductor critical temperature depends upon the chemical structure of superconductor. We anticipate that the superconducting energy gaps for mercury, and lead will show small maxima near 0.21, and 0.11 K, respectively. Anticipated maxima for type I superconductors are due to the two sources of entropy differences between the normal phase and the superconducting phase: 1) dissipative electron scattering in the normal phase; and 2) coherent order in the superconducting phase. The dissipative scattering component of the free energy depends upon the existence of a current, which is often neglected in the study of circuits. In this case the current in the superconducting state is essential to operation of the Meissner Ochsenfeld effect. In some of the elemental superconductors for which high quality H_c data is available, hypothetical curve maxima are anticipated at sub-zero temperatures. The fact basis supporting these details was presented in the late 1950's, and the original authors were careful to point out the significant differences in properties for individual elements.

Transition from normal conductor to superconductor involves an explicit change in entropy for the conductor. The recent experimental demonstration of Landauer's principle, which links thermodynamics and information by Lutz, et al, confirms the need for a change in system conduction free energy at the phase transition between a functioning normal metal conductor and the corresponding superconductor. We submit that this is one of the central questions that must be addressed in superconductivity.

Free energy difference between dissipative and non-dissipative conduction is one of the subjects of this paper. Other subjects include: the connection between this free energy difference and the temperature dependent energy gap in a superconductor; and the contribution of the entropy difference between the bulk normal conductor, which is not a wave mechanically coherent object, and the bulk superconductor, which is a wave mechanically coherent object, to the energy gap of the superconductor. A model for the superconductor energy gap, as reflected in the magnitude of the superconductor critical magnetic field, H_c, is developed. Graphical evaluation of the shape of the energy gap curve with T, for a data series from the literature, is provided in the discussion. Electron free energy differences between dissipative and non-dissipative current flow in metals do not appear to have received much attention in the literature, though non-dissipative superconductors have been known since 1911. Entropy differences between dissipative and non-dissipative electron currents must contribute to the temperature dependent energy gap that characterizes the non-dissipative currents in superconductivity.

This paper points out the relationship between critical magnetic field for a superconductor, $H_c(T)$, and the superconductor energy gap, $\Delta E(T)$, that is an essential feature of the superconducting state. When an external critical magnetic field has been applied to a superconductor, if the external field just matches the internal field, which arises due to the Meissner Ochsenfeld effect, the external field penetrates the superconductor and fills the superconductors energy gap with its energy, proportional to $H_c{}^2$. When the energy of the external magnetic field is equal to the energy gap of the superconductor, Dirac fermions in the superconducting state absorb this energy and bridge the energy gap to a normal conducting state. The Fermi level increases, to its original T_c level. Electrons with basic functions that produce dissipative scattering by Fermi "contact" are available for conduction, and the superconductor quenches.

Theory

Gibbs free energy, equation $G(P, T) = H - TS$, provides the foundation for understanding the possibilities for formation of an energy gap in a superconducting system.

In equation $G(P, T) = H - TS$, G is the Gibbs free energy, H is the enthalpy, and S is the entropy, all at P, and T. The two terms on the right of the equation provide two potential avenues for altering the free energies of systems, like normal conductors or components like conducting electrons. If we focus on the conducting electrons, which are intimately associated with the energy gap in superconductors, their enthalpy and temperature times entropy are the controlling factors. System entropy provides the only source of free energy difference between two phases, like a superconductor and the corresponding normal conductor phase, where the difference in free energy between the two phases is intrinsically temperature dependent.

Electron Enthalpy

It is possible to change electron enthalpy in a conductor by altering electron interactions with other particles particularly electrons and atomic nuclei. This is what happens in chemical bond formation, and in spin pairing Mott transitions, which are known to occur in specific materials including superconductors at low temperatures. In current theory, chemical bond formation is a temperature independent process, mediated entirely by forces due to particle fields. Bond formation includes enthalpy associated with processes including: interactions of charges, columbic attraction and repulsion; and magnetic interactions, magnetic coupling, spin pairing, etc. It corresponds to bonding enthalpy and does not seem a proper candidate for formation of a temperature dependent energy gap in superconductors. In the Mott transition, pairs of single electron conducting states are transformed into pairs of anti-bonding and bonding states, which create a temperature independent energy gap on the insulator side of Mott systems. In this case the energy gap corresponds to the electronic excitation energy associated with the bonding and anti-bonding pair of states. These energy gaps are known to be temperature independent.

Electron Associated Entropy

There are two distinct sources of the energy gap in superconductors both involve entropy differences between the normal and superconducting states of the conductor. One entropy contribution to the electron free energy in superconductors comes from the current in the normal conductor at the phase transition. Entropy arises in normal conductors in the form of resistive electron scattering, which does not occur on the superconductor side of the phase transition. Involvement of electron scattering in resistivity has been known since the late 19th century; however, a purely electronic theory of resistivity that is consistent with the Sommerfeld equation remains to be fully developed. Sommerfeld's equation gives the relationship between thermal conductivity, and electrical conductivity, σ, known as the Wiedemann Franz Law, in terms of the temperature and established constants, equation $\dfrac{K}{\sigma T} = \dfrac{\pi^2}{3}\left(\dfrac{k_B}{e}\right)^2 = 2.44 \bullet 10^{-8} W\Omega K^{-2}$.

Sommerfeld's equation was developed by exclusive use of electron gas wave functions, which indicates that the Born Oppenheimer approximation is functional and lattice wave functions and their components are not involved in the relationship. This specifically means than lattice vibronic quanta, phonons, are not a part of the relationship between electrical conductivity and thermal conductivity in metals, equation $\dfrac{K}{\sigma T} = \dfrac{\pi^2}{3}\left(\dfrac{k_B}{e}\right)^2 = 2.44 \bullet 10^{-8} W\Omega K^{-2}$.

In the normal conductor/superconductor phase transition, the change in entropy at the end of dissipative electron scattering is an exquisitely subtle topic in statistical thermodynamics. The subject can be considered using Landauer's principle, which links information and thermodynamics, as well as standard thermodynamic considerations. The conservative estimate given for the Landauer limit in the measurements reported by Lutz, et al. is ln2 · kT per event, or in this case, per electron. This data provides a credible beginning for understanding the details of the superconductor temperature dependent energy gap, at least at on-set, T_c. Temperature dependence for this contribution to the energy gap is linear. That is, the contribution to the electron free energy will vary as -TΔS. As the temperature decreases the contribution per electron of dissipative electron scattering in the normal phase to electron free energy in the superconducting phase will decrease.

A second contribution to electron free energy in superconductors comes from the entropy difference between the superconducting state, which is coherent, and the normal state, which is non-coherent. At the superconductor phase transition the mechanism for thermal conductivity and heat capacity fundamentally changes. On the superconductor side of the transition all of the conducting electrons are carriers of heat capacity with direct quantum mechanical coupling to the lattice dynamics. The Born Oppenheimer approximation fails here, as does the Sommerfeld equation,

$\frac{K}{\sigma T} = \frac{\pi^2}{3}\left(\frac{k_B}{e}\right)^2 = 2.44 \cdot 10^{-8}\, W\Omega K^{-2}$. On the normal conductor side of the phase transition some of the electrons may not be effective carriers of thermal energy, and Sommerfeld's equation, functions, particularly well at very low temperatures in normal conductor systems.

The shape of the ΔE v. T curves, tells us that the energy gap increases dramatically with decreasing temperature. This increase with decreasing temperature corresponds to the effect of both the increase in current for the internal magnetic field as H_c increases, and the effect of decreasing temperature on the coherence and order in the superconducting state as compared to the normal state in which the Born Oppenheimer approximation applies and there is no coherence of the electronic and vibronic wave functions.

Thermodynamic Model for the Superconducting Energy Gap

When an external magnetic field, H, is applied to a superconductor, it is expelled by the Meissner Ochsenfeld effect. When an external magnetic field penetrates the body of a superconductor, the energy contained in the external magnetic field, proportional to H^2, is absorbed by the superconductor energy gap. When $H = H_c$, the applied magnetic field is sufficiently large to reduce the energy gap to zero, allowing a transition to the normal conducting state. The condition $EH = \Delta E$ enables us to calculate $H_c(T)$. Application of the standard treatment for critical phenomena leads

to equation $\Delta E = a\left(1 - \frac{T}{T_C}\right)^n f(T)$.

In equation above, n is a critical exponent, a is a proportionality constant, and ΔE corresponds to the temperature dependent energy gap of the superconductor. The first factor on the right of equation above corresponds the threshold behavior near the critical temperature, T_c, and f (T) is a slowly varying function normalized to 1 at T = 0. We approximate f(T) by the first order McLaurin expansion, $f(T) = 1 + b\frac{T}{T_C}$.

Since the energy of the uniform magnetic field is proportional to H^2, from equation.

$$\Delta E = a\left(1 - \frac{T}{T_C}\right)^n f(T)$$

we obtain,

$$H_C(T) = H_C(0)\left(1 + b\frac{T}{T_C}\right)^{1/2}\left(1 - \frac{T}{T_C}\right)^{n/2}.$$

A close approximation to equation above is given in equation

$$H_C(T) = H_C(0)\left(1 + \frac{b}{2}\frac{T}{T_C}\right)\left(1 - \frac{T}{T_C}\right)^{n/2}.$$

Equation above has the potential to deal with the variety of critical magnetic curves that are presented by elemental superconductors.

Data Analysis

We have used equation $H_C(T) = H_C(0)\left(1 + \dfrac{b}{2}\dfrac{T}{T_C}\right)\left(1 - \dfrac{T}{T_C}\right)^{n/2}$ to model the critical magnetic fields reported for 10 elemental superconductors for which we found suitable data in the literature. Only zinc, Zn and cadmium, Cd in table have critical temperatures below 1 K. Largely because of the lack of data for low T_c superconductors, the only representatives of the d series superconductors in table I are niobium, Nb, and tantalum, Ta.

Table: Elemental superconductor critical magnetic field v. T data fitting parameters to equation.

$$H_C(T) = H_C(0)\left(1 + \frac{b}{2}\frac{T}{T_C}\right)\left(1 - \frac{T}{T_C}\right)^{n/2}.$$

Elem	$H_c(0)$	unit	b	n	T_c	χ^2	COD
Al	106	G	2.10	2.02	1.20	5.38	0.9995
Zn	50.2	G	2.52	2.13	0.907	1.656	0.9993
Ga	52.6	G	1.43	1.83	1.11	34.2	0.993
Nb	1710	Oe	2.17	1.94	9.22	7639.4	0.9996
Cd	28.9	G	1.73	1.86	0.553	0.312	0.9997
In	294	Oe	1.71	2.03	3.41	0.5965	0.999995
Sn	317	Oe	1.75	2.06	3.73	1.336	0.999996
Ta	640	Oe	1.71	2.03	3.41	0.5965	0.999995
Hg	405	G	2.30	2.03	4.15	3.3569	0.999994
Pb	792	G	2.15	1.96	7.17	9.0923	0.999995

It has been known for more than 50 years that heat capacity of a superconductor, the square of the superconductors critical magnetic field and the superconductor energy gap are all related. The availability of a model makes it possible to enquire about the variance of the energy gap from one element to the next. This assumes that all of the superconductors in table have the same relationship between critical magnetic field energy, $\propto H_c^2$ and the magnitude of the superconductor energy gap.

Figure presents the fitted H_c v. T curves for the 10 superconductors in table. If you look carefully you will see that there are three curves in figure 1 that have maxima at temperatures above-0 K. These elements, Nb, Hg, and Pb all have values for the parameter b greater than their values for the parameter n. Low temperature measurements for elemental mercury, Hg, look like the best

case for evaluating the validity of the model in $H_C(T) = H_C(0)\left(1 + \dfrac{b}{2}\dfrac{T}{T_C}\right)\left(1 - \dfrac{T}{T_C}\right)^{n/2}$ for the critical

magnetic field of a type I superconductor. Using the experimental critical magnetic field data to determine the anticipated maximum in the curve of H_c v. T, we obtained values of 0.21, and 0.11 K for the anticipated maxima in H_c for mercury and lead respectively.

From the analytical details presented in the critical magnetic field studies of the Mapother research group it is clear that as a group they were aware of the regularities of the curve shapes in the critical magnetic field data for elemental superconductors. Historically the equation used to model the temperature dependence of the critical magnetic field is shown as equation $H_C(T) = H_C(0)\left(1 - \dfrac{T^2}{T_c^2}\right)$.

Mapother's group modeled the real H_c data using an equation for deviations from equation above. In their analysis mercury and lead gave positive deviations and the remainder of the elemental superconductors, for which H_c data was available gave negative deviations. These two groups of elements correspond one to one to elements for which parameter b in table is greater than n (positive deviation group) and parameter b is less than the corresponding n (negative deviation group). For the type I superconductor cases in table for which the literature provides extensive high quality data sets the coefficient of determination, COD in table using equation

$$H_C(T) = H_C(0)\left(1 + \frac{b}{2}\frac{T}{T_C}\right)\left(1 - \frac{T}{T_C}\right)^{n/2}, \text{ is essentially at the maximum.}$$

Using equation above for these elements, there is no need to model the deviations.

Niobium's large, relatively recent, H_c data set, shows significant scatter using equation above ((1-COD)>4·10⁻⁴). It seems likely that this is a consequence of the nearby second order phase transition to the magnetic vortex lattice, which has a different magnetic description than a type I magnetic field. We do not have confidence in the use of equation above to fit data for type II superconductors. The curve is simply for comparison.

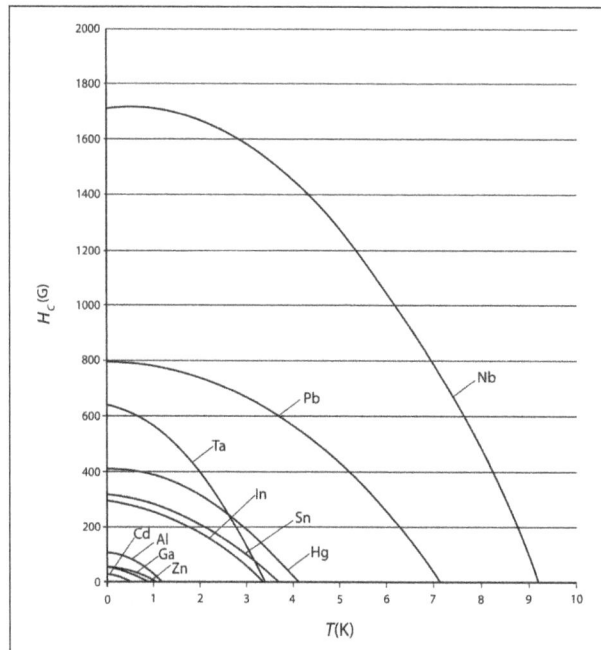

Critical magnetic field curves to experimental data for superconductors in table.

Lines showing the fit of the square of critical magnetic field v. T from equation

$$H_C(T) = H_C(0)\left(1 + \frac{b}{2}\frac{T}{T_C}\right)\left(1 - \frac{T}{T_C}\right)^{n/2}, \text{ to the square of experimental data, points, for the elements}$$

listed in table are presented in figures below. Figure is devoted to the d series elements, niobium and tantalum. Figures below present superconducting elements at one bar from the main group in the periodic table and column 12. Main group and column 12 elements with T_c above 1.5 K are presented in figure and those with T_c below 1.5 K are in figure. The scaling factors used to reduce the scale in the three figures for the y-axis values are 10^4, 10^3 and 10^2 for figures below respectively.

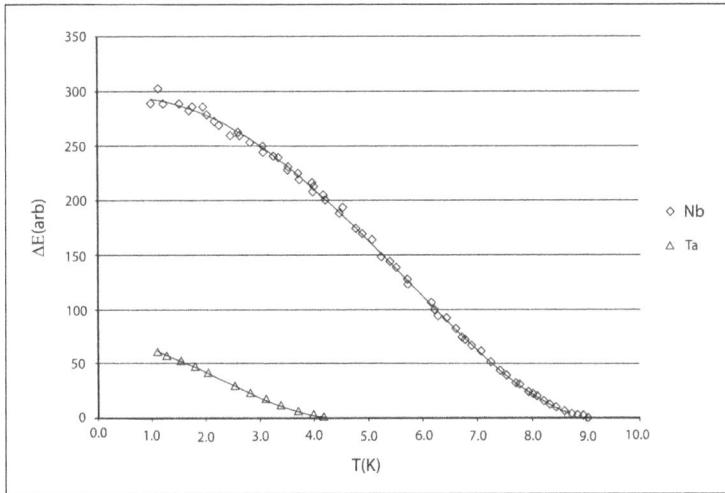

Relative energy gap (arbitrary units) v. T (K) for niobium, Nb[16] and tantalum, Ta[17]. Points show the square of experimental critical magnetic field divided by 10^4. Solid line from values obtained

by equation $H_C(T) = H_C(0)\left(1 + \dfrac{b}{2}\dfrac{T}{T_C}\right)\left(1 - \dfrac{T}{T_C}\right)^{n/2}$, using parameters in table on the same scale as the points.

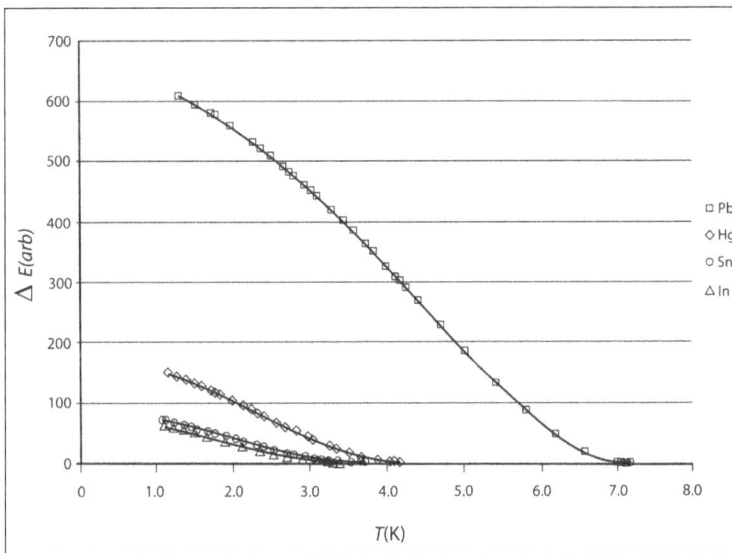

Points show the square of experimental critical magnetic field divided by 10^3. Solid line from values obtained by equation $H_C(T) = H_C(0)\left(1 + \dfrac{b}{2}\dfrac{T}{T_C}\right)\left(1 - \dfrac{T}{T_C}\right)^{n/2}$, using parameters in table on the same scale as the points.

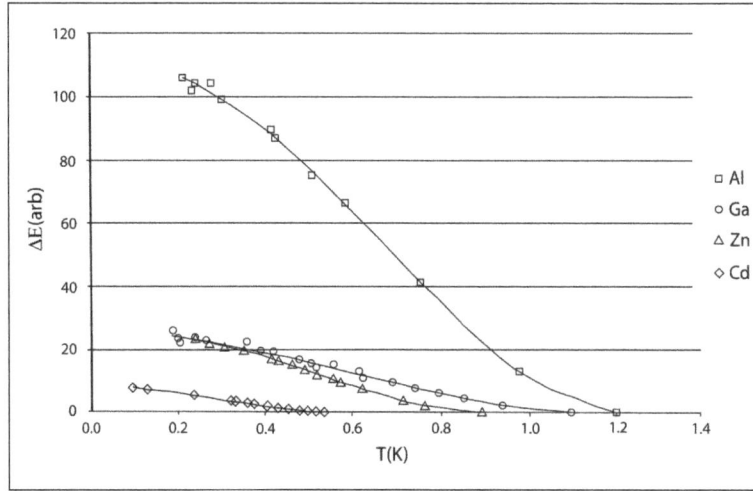

Points show the square of experimental critical magnetic field divided by 10^2 Solid line from squares of values obtained by equation $H_C(T) = H_C(0)\left(1 + \dfrac{b}{2}\dfrac{T}{T_C}\right)\left(1 - \dfrac{T}{T_C}\right)^{n/2}$, using parameters in table on the same scale as the points.

Above four figures illustrate the variability in the appearance of relative critical magnetic field energies and the associated superconductor energy gaps for the elements listed in table. To illustrate the variance in the shapes of the superconductor, energy gap or critical magnetic energy, curves with temperature we have evaluated the relative energy gap, equation $\Delta E_{SC}(T) \simeq gH_C(T)^2$, at both 15% and 90% of the critical temperature.

In equation above, $\Delta E_{sc}(T)$ is the superconductor energy gap at the critical magnetic field as a function of temperature. For the data taken from the sources in table, the scaling factor, g, is unknown, and is arbitrarily set to 1 for all relative gap comparisons for elements in table. External magnetic fields have nothing to do with the creation of an energy gap in superconductors. The superconductor energy gap arises from the Gibbs free energy difference for the conducting electrons between the normal conductor and the superconductor. When the energy of the external magnetic field is equal to the energy gap of the superconductor, electrons in the superconducting state are able to absorb this energy and bridge the energy gap to a normal conducting state resulting in a dissipative resistance initiated quench for the superconductor. The dissipative resistance arises from the participation of electrons in basis states that support Fermi "contact" dissipative scattering in the normal conducting state. This energy gap is the maximum value for the energy gap at that temperature, which varies with the internal magnetic field of the superconductor.

The ratio of the square of the fitted values of critical magnetic field at $0.15 \ast T_c$ to those at $0.9 \ast T_c$ are shown in table II along with the ratio of extrapolated values of $\Delta E(0)$ to T_c.

There are a large enough number of elements in table II to permit the conclusion that use of a single value for energy gap at $T \approx 0$ will not be a reasonable approximation to explain the experimentally based results.

The average of the $\Delta E._{15Tc}/\Delta E._{9Tc}$ ration of magnetic energies in Table is 26.2 with a standard deviation of ±4.87. This standard deviation is less than 20% of the average value for the ratios. The average value of the energy gap, estimated using critical magnetic fields for these 10 superconductors with its standard deviation was: 6.23E+4 (arb.)±9.78E+4 In this case the standard deviation is fully 1.5 times the average of the magnitude of the energy gaps divided by the respective T_cs.

Table: Energy gap ratio for $0.15T_c$ to $0.9T_c$ and $\Delta E(0)/T_c$ for elements in table.

Element	$\Delta E._{.15Tc}/\Delta E_{.9Tc}$	$\Delta E(0)/T_c$ (arb.)
aluminum	26.7	9.36E+03
zinc	32.2	2.78E+03
gallium	23.1	2.49E+03
niobium	22.1	3.17E+05
cadmium	23.4	1.51E+03
indium	19.2	2.53E+04
tin	33.4	2.69E+04
tantalum	31.5	1.20E+05
mercury	25.8	3.94E+04
lead	22.5	8.77E+04
average ± std dev	26.0±4.86	6.32E+4±9.75E+4

The experimentally supported observations documented in above table, do not support a model that calls for all type I superconducting elements having the same value, e.g., 3.528 $k_B T_c$, for $\Delta E(0)/T_c$ or a closely related ratio. Table above was constructed by preparing a spreadsheet using

the parameters in the previous table, and equation $H_C(T) = H_C(0)\left(1 + \dfrac{b}{2}\dfrac{T}{T_C}\right)\left(1 - \dfrac{T}{T_C}\right)^{n/2}$, to model

the critical magnetic fields of the 10 elemental superconductors for which we have literature data. The temperature resolution used was 0.01 K. Values from the fitted curves were then used with equation $\Delta E_{SC}(T) \simeq gH_C(T)^2$, to construct above table. For the purposes of this exercise oersteds, Oe, were considered equivalent to gauss, G.

There are five main group metals in the above table. These are elements in periodic table columns 13 and 14. Values for the ratio, $\Delta E_{.15Tc}/\Delta E_{.9Tc}$, in table for these elements range from 19.2 to 33.4 for indium and tin respectively. The wide variation in this ratio suggests chemical origins for the observed differences. These differences, like the pattern of critical temperatures for the one bar superconductors in the periodic table, table below, and must have a chemical origin. A probable origin for this pattern is in the density of molecular electronic states in the superconductor.

Understanding critical temperatures in elemental superconductors, requires knowing that conducting electrons in s basis conduction bands will have a partial wave scattering cross section as a function of temperature down to the lowest attainable temperatures. Furthermore, for s basis wave functions the threshold cross section is independent of temperature, so this scattering will continue at the lowest attainable temperatures. Partial wave generated electron scattering will effectively stop any phase transition for a metal to superconductivity. Partial wave scattering at threshold exponentially drops with temperature as the 4l power for conducting electrons in wave functions with basis l>0.

Table: (Color online) Periodic table showing elemental bulk superconductors at one atmosphere with their critical temperatures, T_c.

Legend:
- Symbol → Nb; Atomic Number → 41; Critical Temperature (T_c) in Kelvin (K) → 9.25
- $T_c \geq 1.5$ K
- $1.5 > T_c > 0.1$ K
- $T_c \leq 0.1$ K

Period	1	2	3	4	5	6	7	8	9	10	11	12	13	14	15	16	17	18
1	H 1																	He 2
2	Li 3	Be 4 0.026											B 5	C 6	N 7	O 8	F 9	Ne 10
3	Na 11	Mg 12											Al 13 1.18	Si 14	P 15	S 16	Cl 17	Ar 18
4	K 19	Ca 20	Sc 21	Ti 22 0.5	V 23 5.4	Cr 24	Mn 25	Fe 26	Co 27	Ni 28	Cu 29	Zn 30 0.85	Ga 31 1.08	Ge 32	As 33	Se 34	Br 35	Kr 36
5	Rb 37	Sr 38	Y 39	Zr 40 0.6	Nb 41 9.25	Mo 42 0.92	Tc 43 8.2	Ru 44 0.5	Rh 45	Pd 46	Ag 47	Cd 48 0.57	In 49 3.4	Sn 50 3.7	Sb 51	Te 52	I 53	Xe 54
6	Cs 55	Ba 56	La 57 6.0	Hf 72 0.38	Ta 73 4.4	W 74 0.01	Re 75 1.7	Os 76 0.7	Ir 77 0.1	Pt 78	Au 79	Hg 80 4.15	Tl 81 2.4	Pb 82 7.2	Bi 83	Po 84	At 85	Rn 86

In developing an understanding of the chemical origins of the pattern of critical temperatures for the one bar superconductors in the periodic table, we did two things: 1) used all the data for the one bar superconductors above period 7; and 2) focused on the magnitude of T_c for each element. The relatively high T_c superconductors in the d series, superconductors in columns 3 through 9 in periodic table, are in odd numbered columns. This suggested that the Fermi level for the superconductors in columns 3, 5 and 7 of periodic table has a closed valence level s sub-shell for the metallic state of these conductors at very low temperatures. For the main group superconductors in periodic table, it is possible for superconductors in columns 13 and 14 to be conductors with a closed s sub-shell at low temperatures. This effect is classically known as the "inert pair" effect, in chemistry. Having a closed s sub-shell would make these conductors p wave superconductors at temperatures below T_c.

Mercury's relatively high T_c of 4.15 K seems likely to be the result of the fact that metallic mercury can form two electron 6s bonds in the solid state that leave low temperature metallic mercury as an effectively closed s sub-shell element. Zinc and cadmium, in the same column of the periodic table as mercury, are not known to form strong s-s single bonds as are known for mercurous salts like calomel, Hg_2Cl_2, no "inert pair" effect.

The possibility of using the same strategy for examining the chemical basis for variance in the elemental superconductor energy gap that was used for critical temperatures is seriously limited by the relatively small amount of data that is available on critical magnetic fields and/or direct measurements of energy gaps, as compared to critical temperatures which are known for all elemental superconductors. Since the problem of the dispersion of values for the magnitude of the energy gap at near zero temperatures in superconductors is a problem on the chemical side of chemical physics, it is hoped that this problem will attract the attention of condensed matter scientists with a chemical background.

In establishment of the current paradigm for understanding superconductivity it appears that a very limited sample of superconductors were used to form the conclusion that the superconductor energy gap at-0 K divided by the critical temperature for the superconductor was a constant for temperatures near 0 K that did not vary from one superconductor to the next.

References

- Superconductor-2699012: thoughtco.com, Retrieved 18 June, 2019

- What-is-superconductivity: electrical4u.com, Retrieved 15 February, 2019

- Type-i-and-type-ii-superconductors: winnerscience.com, Retrieved 04 May, 2019

- Magnetization-function-type-I-superconductor-magnetic-field-2405-004-1C47E03D: cdn.britannica.com, Retrieved 17 August, 2019

- "Superconductivity Revisited" (CRC Press, Boca Raton, in press, 2012)

Materials-based Superconductors

There are various special material-based superconductors. These include A-15 compounds, $CeRu_2$, pyrochlore oxides, rutheno-cuprates, magnetic superconductors or chevrel phases, heavy fermion superconductors, oxide superconductors without copper, etc. This chapter closely examines these material-based superconductors to provide an extensive understanding of the subject.

A-15 Compounds

Chemical compounds of binary composition A_3-xB_{1+x} crystallize into many different structures, depending on the value of x, temperature and pressure. One of the structures existing near x = 0, A_3B (where A = Nb, V, Ta, Zr and B = Sn, Ge, Al, Ga, Si) has the structure of beta-tungsten, designated in crystallography by the symbol A-15, and is superconducting. Hardy and Hulm first discovered A-15 superconductor (V_3Si) in 1954. Intermetallic compound Nb_3Ge has a critical temperature T_c = 23.2 K, while Nb_3Ga shows T_c = 20.7 K, Nb_3Al, T_c = 19.1 K and Nb_3Sn, T_c = 18.3 K. Figure shows the structure of the binary A_3B compound. The transition temperatures of several A – 15 structures are given in table.

The critical temperature and second critical field increase in Nb_3Al compound, adding Ge and Cu in Nb/Al at the initial stage of the rapid heating, quenching and transforming (RQHT) process, and they attain for Nb_3Al-Ge(T_c = 19.4 K, H_{c2} = 39.5 T) and for Nb_3Al-Cu(T_c = 18.2 K, H_{c2} = 28.7 T). Moreover, the superconductors Nb_3Al-Ge, Cu have highest critical current densities among all metallic multifilamentary superconductors at H > 20 T and T = 4.2 K. Only significantly more expensive HTSC, based on Bi-2223 and Bi-2212, have the near values of supercurrent at T < 20 K.

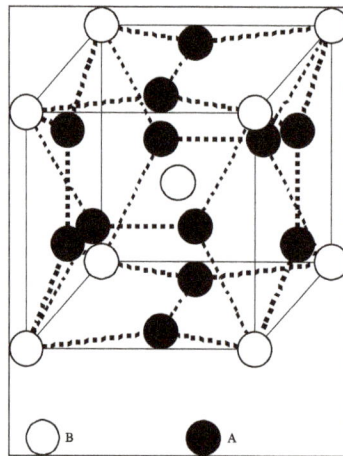

Crystalline structure of A_3B compound (A – 15 superconductors). The atoms A form one-dimensional chains on each face of the cube. Chains on the opposite faces are parallel, while on the neighboring faces they are orthogonal to each other.

Table: The critical temperatures, $T_c(K)$, of A-15 compounds.

Ti_3Sb	6.5	Ti_3Ir	4.2
$Zr_{80}Sn_{20}$ q	0.92	Ti_3Pt	0.5
Zr-Pb	0.76	$V_{29}Re_{71}$	8.4
$Zr_{\sim3}Bi$ p	3.4	$V_{50}Os_{50}$	5.7
V-Al f	11.8	$V_{65}Rh_{35}$	≈ 1
V_3Ga	15.9	$V_{63}Ir_{37}$	1.7
V_3Si	17.0	$V_{\sim3}Pd$	0.08
$V_{\sim3}Ge$	6.0–7.5	V_3Pt	3.7
$V_{\sim3}Ge$ f	6.0–11.0	$Nb_{75}Os2_5$	1.0
$V_{\sim79}Sn_{\sim21}$	4.3	$Nb_{75}Rh_{25}$	2.6
V-Sn q	7.0–17.0	$Nb_{72}Ir_{28}$	3.2
$V_{77}As_{23}$	0.2	Nb_3Pt	11.0
$V_{76}Sb_{24}$	0.8	$Ta_{85}Pt_{15}$	0.4
Nb_3Al	19.1	$Cr_{72}Ru_{28}$	3.4
Nb_3Be	10.0	$Cr_{73}Os_{27}$	4.7
Nb_3Ga	20.7	$Cr_{78}Rh_{22}$	0.07
Nb_3Pb	5.6	$Cr_{82}Ir_{18}$	0.75
$Nb_{\sim3}In$ p	8.0–9.2	$Mo_{40}Tc_{60}$	13.4
$Nb_{82}Si_8$ q	4.4	$Mo_{\sim65}Re\sim_{35}$ f	≈ 15 (A-15)
Nb-Si f	9.3	$Mo_{75}Os_{25}$	12.7
Nb-Si f	4.0–8.0	$Mo_{78}Ir_{22}$	8.5
Nb-Si f	11.0–17.0	$Mo_{82}Pt_{18}$	4.6
Nb-Ge q	6.0–17.0	$W_{\sim60}Re_{\sim40}$ f	11.0
Nb-Ge f	23.2	$Ta_{\sim80}Au_{20}$	0.55
Nb_3Sn	18.3	Zr_3Au	0.9
Nb-Sb	2.0	$V_{76}Au_{24}$	3.0
$Nb_{\sim3}Bi$ p	3.0	$Nb_{\sim3}Au$	11.5
$Ta_{\sim3}Ge$ f	8.0		
$Ta_{\sim3}Sn$	8.3		
$Ta_{\sim3}Sb$	0.7		
Mo_3Al	0.58		
Mo_3Ga	0.76		
$Mo_{77}Si_{23}$	1.7		
$Mo_{77}Ge_{23}$	1.8		

Gap Structures of A-15 Alloys from the Superconducting and Normal-State Break-Junction Tunnelling

A well-known A-15 compound Nb_3Sn was investigated by the break-junction tunnelling technique with high superconducting critical temperature $T_c \approx 18$ K. Relevant energy-gap values were measured at T = 4. 2 K and these manifested as conductance peaks at bias voltages $2\Delta/e$ = 4–6 mV. Here, T is temperature and e> 0 is the elementary charge. In addition to superconductivity-driven gap structures, reproducible humps were also detected at biases ±20–30 mV and ±50–60 mV for T = 4.2 K. Such hump features, complementary to coherent peaks at the superconducting-gap edges, apparently resemble the pseudo-gap manifestations inherent to high-T_c superconductors. These humps remain the only gap-related features above T_c. Possible origins of these structures are discussed with emphasis on the charge–density–wave (CDW) formation. CDWs are accompanied by periodic lattice distortions and are related to the structural phase transition discovered decades ago in Nb_3Sn. The current-voltage characteristics exhibit asymmetries, being probably a consequence of normal-metal junction shores or due to the vanishing symmetry of the junction conductance when CDWs are present in both electrodes.

Materials with A-15 (β-tungsten) crystal structure possessed the highest superconducting critical temperatures, T_c, over a long period of time until Cu-based superconducting ceramic oxides were discovered. The compound Nb_3Sn is a representative material with $T_c \approx 18$ K among those inter-metallic alloy compounds and possesses stable metallurgical characteristics. Therefore, Nb_3Sn and some other older and newer materials, such as MgB_2 and Fe-based superconductors, are competitive with cup rates in technological applications of superconductivity.

The crystal structure of Nb_3Sn (as well as of its A-15 relatives) is not a layered one as that of the copper-oxide or Fe-based high T_c superconductors. Instead, it includes orthogonal linear chains of Nb atoms along each principal cube axis direction. This quasi-one-dimensional feature served as a guide for the Labbe–Friedel model, which predicted structural anomalies driven by peculiarities of the normal-state electron density of states, N(E), near the Fermi level (a cooperative Jahn–Teller effect). More involved Gor'kov model, taking into account inter alia the inter-chain correlations, also leads to the N(E) singularity and a Peierls phase transition. Whatever the theoretical details, the primordial (high-temperature, high-T) electron spectrum in Nb_3Sn is unstable towards low-T phase transformation.

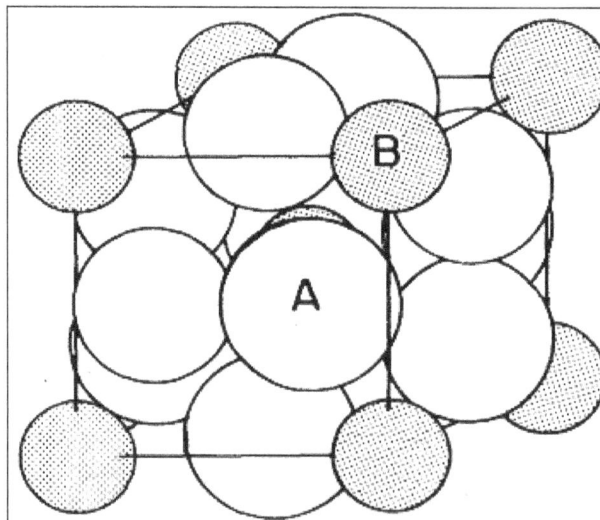

The A-15 crystal structure for the compound formula A_3B. In our case, A stands for Nb whereas B stands for Sn.

In fact, A-15 cubic compounds, including Nb_3Sn, are well known to undergo tetragonal distortion below a certain temperature Tm, being higher than T_c. Hence, some part of the Fermi surface is gapped by CDWs, which is detrimental to superconductivity, because of the reduced electron density of states, which to a large extent determines T_c. The interplay between periodic lattice distortions accompanied by CDWs in the electron subsystem, on the one hand, and superconductivity, on the other hand, was intensively studied both experimentally and theoretically and applied to Nb_3Sn as a particular case.

The contest between superconducting and CDW (dielectric) order parameters should inevitably lead to a superposition of the corresponding energy gaps in the overall electron spectrum modified by both Cooper and electron–hole pairings. Any experimental technique measuring the gapped electron density of states below T_c or the convolution with its counterpart (if any) should be sensitive to the interplay between phenomena. A typical example of such a method is electronic quasi-particle tunnelling spectroscopy. Moreover, if multiple gaps of the same nature (usually, those are recognized as superconducting gaps) are inherent to the reconstructed electron spectrum or multiple gaps are generated by some kind of proximity effect, they would influence the tunnel conductance together. Multiple superconducting gapping in Nb_3Sn was also suggested on the basis of heat capacity and point-contact conductivity measurements. On the contrary, subsequent heat capacity studies were considered as manifestations of a single superconducting gap.

The origin of superconductivity and the gapping features in A-15 compounds have been explored by various methods. Namely, fabrication of corresponding tunnel junctions and electron tunnelling spectroscopic measurements were intensively carried out to probe the electron–phonon interaction in terms of the Fröhlich–Eliashberg function $\alpha^2 F(\omega)$ and to extract the superconducting-gap values 2Δ. Those junctions made for tunnelling spectroscopy purposes included artificial oxide barriers that might cause spurious features in the tunnelling spectra. On the other hand, pristine junctions of Nb_3Sn samples such as break junctions or cleaner direct contacts turned out to be of better quality so that the ambiguous influence of the barrier was avoided both in the tunnel and point-contact conductivity regimes. Those measurements revealed a single clear-cut superconducting gap and another feature at higher voltages most probably connected with the structural transition. Tunnelling measurements of Nb_3Sn single crystals are presented using a break-junction technique that has been improved. The experiments were carried out focusing on both the superconducting-gap characteristics and the electronic peculiarities of N(E) emerging due to the structural (martensitic) transition intimately associated with CDWs. The latter could be due to the Peierls or excitonic (Coulomb) transitions of the parent high-T state. In the mean-field approach, the coupled system of equations describing competing superconducting and electron–hole pairings is the same for any microscopic picture of the CDW pairing. Therefore, the consequences important for our subsequent analysis are similar for both kinds of electron–hole instabilities at this semi-phenomenological level, although one should bear in mind the necessity of the microscopic justification for any adopted model. According to earlier findings in the point-contact measurements, we expected the manifestations of CDWs also in the quasi-particle conductance G(V) = dI/dV, where I denotes the quasi-particle tunnel current across the break junction. The quantity G(V) in CDW superconductors is a complicated functional of the superconducting, Δ, and dielectric (CDW), Σ, energy gaps being no more a simple convolution of the electron density of states as in the conventional Bardeen–Cooper–Schrieffer (BCS) model of superconductivity. From the experimental point of view, tunnel conductances G(V) distorted by CDW gapping are well known for a number of relevant materials.

Experimental Procedures

Nb_3Sn single crystal samples were grown by a standard vapour transport method. The temperature dependence of the electrical resistance for an Nb_3Sn crystal is shown in figure. It was measured using the break-junction configuration just before the breaking. To avoid any difficulty in forming a clean junction interface on the small surface area of the tiny crystalline piece, making such a break junction is the best method. A cryogenic fracture at 4.2 K of the crystal piece provides the crucial advantage of this technique. The fracture is performed by at first mounting the sample on the flexible substrate with four electrodes, subsequently stressed by an external bending force. Thus, a fresh and clean junction interface appears that can provide the undistorted gap features both for superconducting and semiconducting electron spectra including extremely surface-sensitive delicate compounds. This junction design exhibits the symmetric superconductor–insulator–superconductor (SIS) geometry.

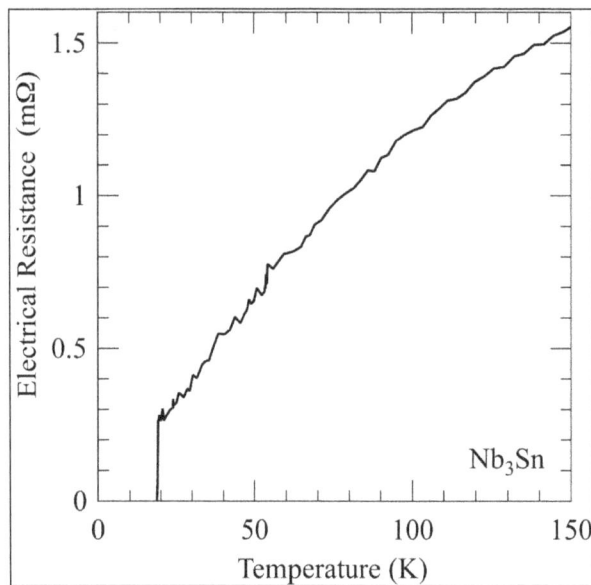

The temperature dependence of the electrical resistance, R (T), for Nb_3Sn used in the present break-junction measurements.

CeRu₂

The interplay between magnetism and superconductivity engulfs two of the richest areas of solid state physics. The coexistence of the two phenomena was first extensively studied in the Chevrel phases where in certain pure compounds magnetic order appears below the superconducting transition temperature, T_c. For these materials the magnetically ordered ions are only weakly coupled to the conduction electrons. A different situation pertains for some of the heavy fermion superconductors such as $U_{1-x}Th_xBe_{13}$, UPt_3, and URu_2Si_2. The small size of the ordered magnetic moments in these compounds relative to their Néel temperatures attests to the more complex many-body origins of their magnetism. It is remarkable that the latter two compounds have magnetic ordering temperatures roughly an order of magnitude higher than T_c and that the magnetic order persists into the superconducting state. In $CeCu_2Si_2$, the most studied Ce-based heavy fermion superconductor, the magnetism is relatively strong and in competition with superconductivity rather than coexisting with it.

Our finding is that the cubic Laves phase superconductor $CeRu_2$ condenses into a static magnetic state at a temperature $T_M \simeq 40$ K which persists into the superconducting state below $T_c = 6.1$ K. The evidence comes from both muon-spin-relaxation (μSR) measurements and ac susceptibility measurements on a single crystal. Our work supports the interpretation that anomalies seen in recently presented high field measurements are due to the occurrence of static magnetism at T_M.

In the superconducting state of $CeRu_2$ an abrupt transition from irreversible magnetic behavior near the upper critical field to almost perfectly reversible behavior at lower fields occurs. The robustness and well-defined nature of this transition has led to the contention that it might be due to some underlying transition within the superconducting state, rather than due to a continuous evolution of flux pinning effects alone. In a recent neutron study the correlation length of the flux line lattice was measured. When interpreted within a theory of weak collective pinning, pin spacing of the order of the superconducting coherence length was deduced. So far no evidence as to the physical origin of the pinning mechanism has been forthcoming. In this light, the existence of magnetic order raises the possibility that the pinning is magnetic in origin and not necessarily related to crystalline defects.

Early studies concerning the coexistence of magnetism and superconductivity related to $CeRu_2$ considered compounds where the Ce had been partially substituted by a third ionic species. It was found that the replacement of Ce with significant quantities of other lanthanide metals can give rise to short-range ferromagnetic correlations. On substituting higher concentrations of these elements the superconductivity is eventually destroyed and replaced by long-range ferromagnetic order. These results should not be confused with the data presented in this article, where we examine only the pure un-substituted compound. In the pure compound the transition is indeed quite subtle and explains why it was not picked up in previous dc magnetization studies. As in a previous investigation18 we do not resolve any anomaly in the resistivity near T_M.

The μSR sample was a disk of ~25 mm diameter and ~0.5 mm thickness, comprising of a mosaic of slices glued on a 5N silver plate (40×40 mm^2). These slices were cut from a large grain polycrystalline ingot of $CeRu_2$. The single crystal used in the susceptibility study was grown by the Czochralski method and had a mass of 1.7 g. No second phases were detectable in similarly prepared crystals in both electron microprobe and high resolution electron microscope studies. The residual resistivities of similarly prepared crystals are of the order 10 $\mu\Omega$ cm.

The μSR measurements were performed with the MuSR spectrometer at the ISIS surface muon beam facility. The spectra were recorded with a closed cycle refrigerator for temperatures between 21 and 151 K and with a helium ("Orange") cryostat for low temperatures down to 2.8 K. Some cross checked spectra were recorded at temperatures up to 49 K with the helium cryostat. The ac susceptibility was measured by the usual inductive technique with a driving field of 3.5 mT at 35 Hz. The crystal was oriented with a low symmetry direction parallel to the ac field for geometric convenience since any magnetic anisotropy is expected to be insignificant ($CeRu_2$ is cubic).

The basic physical quantity measured in our μSR experiment is the muon-spin depolarization function $P_z(t)$ which is simply related to the distribution of fields experienced at the muon stopping site. The measurements correspond to a longitudinal geometry, in which the muon beam polarization is parallel to the incident beam (Z axis) and the positron detectors. We have carried out measurements in zero field and with an external applied field of 1 mT (parallel to Z). The residual magnetic field on the sample during the zero field measurements was $\leq 1 \ \mu$ T.

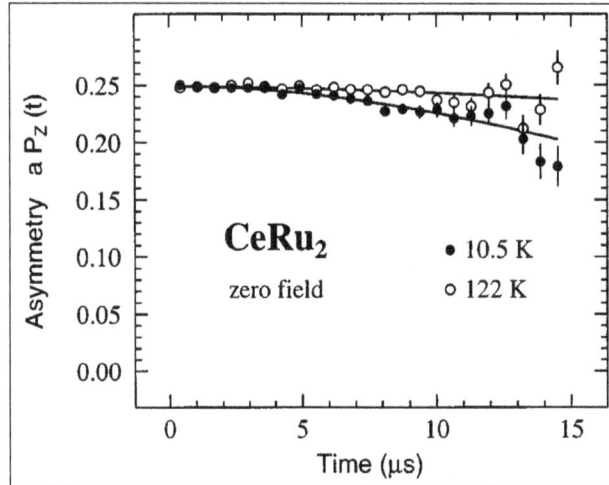

Typical zero field spectra recorded on CeRu$_2$ at 10.5 and 122 K. The lines are fits to the sum of a parabolic depolarization function and a constant term. The relaxation rate is clearly stronger at low temperature.

In figure we present typical zero field spectra. All the spectra are well analyzed by the function,

$$aP_Z\left(t\right) = a_s P_s\left(t\right) + a_{bg}.$$

P_s(t) describes the relaxation due to the sample and the second term in equation above accounts for the muons stopped in the sample holder, cryostat walls and windows. By definition $P_Z\left(0\right) = P_s\left(0\right) = 1$. Measurements at zero field with only the silver plate and no sample showed that the second component does not relax. The data are well described by $P_s\left(t\right) = 1 - \Delta^2 t^2$. The parabolic character of the spectra is clearly seen in figure. A transverse field measurement in the superconducting phase allowed us to determine $a_{bg} : a_{bg} = 0.051$. This a$_{bg}$ value was used as a fixed parameter in the fit. As is then found to be constant over the temperature range investigated a$_s = 0.198$.

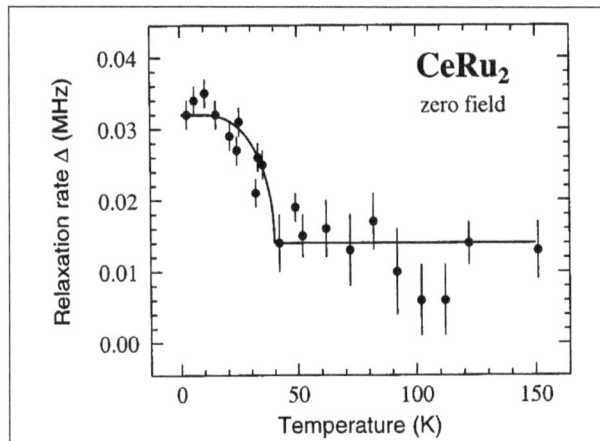

Temperature dependence of the Gaussian muon-spin-relaxation rate, Δ, in CeRu$_2$ at zero field. The line is the Brillouin function prediction for a spin S = 1/2 and T$_M$ = 40 K. This result provides evidence for the occurrence of static electronic magnetism at TM $\simeq$ 40 K.

In figure we display D versus the temperature. While at high temperatures Δ is roughly temperature independent with a value of ~ 0.014 MHz, it increases sharply below $T_M \sim 40$ K to a value of ~ 0.032 MHz at low temperature. Superconductivity does not seem to influence the relaxation rate. That $P_s(t)$ is quadratic in time is a strong indication that the muons are stationary and their spin depolarized by either a static field distribution or a very small coherent field at the muon site. This interpretation is confirmed by the measurements at 1 mT, which show that the depolarization of the muon spin is supressed at both low and high temperature.

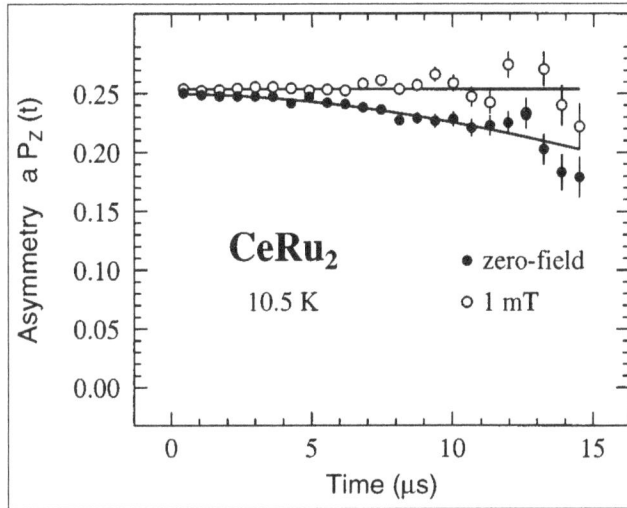

Comparison between zero field and longitudinal field spectra recorded on CeRu$_2$ at 10.5 K. The fact that the depolarization is suppressed by an applied longitudinal magnetic field is an additional proof that the field distribution at the muon site is static.

As a first step to interpret the data of figure, given a stationary muon, we calculate the relaxation at high temperatures induced by the nuclear magnetic moments (uniquely carried by the [99]Ru and [101]Ru nuclei of abundance 12.8% and 17%, respectively). Such a depolarization mechanism would indeed give a parabolic form for $P_s(t)$ for which Δ is then identified with the Kubo-Toyabe relaxation rate due to the nuclear moments, $\Delta_{KT,n}$. There are three possible interstitial muon stopping sites in the cubic Laves phase structure, denoted 2-2, 3-1, and 4-0 where the first (second) digit denotes the number of nearest-neighbor ruthenium (cerium) atoms. For the lattice parameter a=7.538 Å, and neglecting the electric field gradient (EFG) acting on the Ru atoms due to the muon and the lattice environment (the Ru atoms are not in a site of cubic symmetry) we find $\Delta_{KT,n}$ = 0.042, 0.056, and 0.070 MHz for the three sites, respectively. None of these values can explain the measured small damping rate. The difficulty encountered to explain the measured $\Delta_{KT,n}$ at high temperatures is not new. The 2-2 site which has been deduced for the isostructural compound CeAl$_2$ from transverse field measurements does not explain its zero field spectra: the observed $\Delta_{KT,n}$ is again much smaller than given by the simple calculation. These difficulties are probably all related to the neglect of the EFG in the calculation. An alternative possibility is that the muons stop at atomic voids. In this case we cannot compute $\Delta_{KT,n}$ reliably because the position of the atoms are then drastically changed relative to the unperturbed lattice.

Having ruled out the possibility of a mobile muon, the increase of the relaxation rate below T_M must result from the appearance of a very small coherent magnetic field or a broadening of the field distribution which can be either of nuclear or electronic origin. A nuclear origin for the broadening can be eliminated since it would require an unreasonable change of the crystal lattice that has not been detected: a lattice contraction of ~ 25% is needed to explain the fractional change in D with temperature. Therefore the additional relaxation rate detected at low temperatures must be due to magnetic moments of electronic origin. This interpretation is strongly reinforced by the low field magnetization measurements presented below and the high field data of Nakama et al. which are consistent with a magnetic transition at T_M.

Because of the extremely small value of the relaxation rate, the parabolic shape of $P_s(t)$ is a limiting form of either the Kubo-Toyabe function or of an extremely low frequency oscillating signal. The Kubo-Toyabe depolarization function corresponds to a Gaussian field distribution of width Δe (in frequency units, $\Delta_e^2 = \Delta^2 - \Delta_{KT,n}^2$) at the muon site which characterizes a spatially disordered or incommensurate magnetic state, whereas a low frequency oscillating signal is the signature of a coherent magnetic structure with an appreciable correlation length and small magnetic moments. The μSR data cannot distinguish between these possibilities. Under the assumption that the muon spin is depolarized by a field distribution and senses only the dipolar fields from the electronic magnetic moments localized on the Ce atoms, we estimate the Ce magnetic moment: $\mu_{Ce} \gtrsim 10^{-4} \mu_B$. Assuming equal moments on both Ce and Ru sites, as suggested by a recent polarized neutron study, we find about $10^{-4} \mu B$. If we suppose that the increase in damping is in fact due to the appearance of a coherent magnetic field at the muonsite, this field would be 0.05 mT. This corresponds to a μ_{Ce} of the same range as previously estimated. These are the smallest values of electronic moments ever detected. They have however been derived using a simple localized magnetic model. In view of their extremely small value, a bandlike model is probably more appropriate.

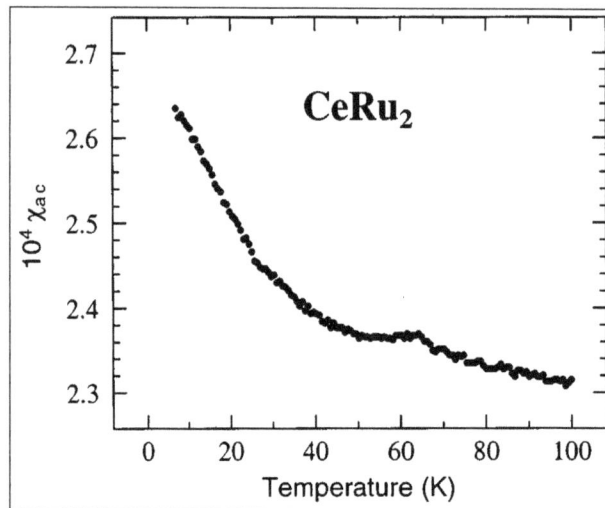

Temperature dependence of the real part of the ac susceptibility of a crystal of $CeRu_2$. In SI units, the susceptibility is dimensionless. The measurements were made for increasing temperature after initially cooling the sample to just above T_c. We observe a plateau followed by an increase at ~ $T_M = 40$ K, the ordering temperature deduced in the μSR experiment.

We have analyzed our spectra supposing that the small detected moment is uniformly distributed in the sample. Another possibility that might be considered is that the depolarization is caused by only a small volume fraction of the sample. From the magnitude of the depolarization at 14.5 μ s we can conclude that at least 15 volume % of the sample is responsible for the depolarization.

A magnetic moment greater than $\approx 10^{-3}\,\mu B$ would be inconsistent with the observed quadratic shape of the depolarization. We note that such a large fraction of any second phase was not detected in our sample.

The ac susceptibility data displayed in figure shows a plateau starting at ~ 60 K followed by a strong increase below T_M. We do not have a definite explanation for the occurrence of the plateau, but the accumulated evidence for a weak magnetic signal in $CeRu_2$ at T_M, from our zero field μ SR and low field susceptibility measurements as well as the high field results from Nakama et al., points definitively to the occurrence of a magnetic transition at T_M. In order to better characterize the magnetic state we have also carried out some measurements in low field with commercial dc superconducting quantum interference device magnetometers, in particular to test for the possible occurrence of magnetic hysterisis. Within our experimental uncertainties we fail to find any such effects.

Small static moments, but still larger by an order of magnitude, have been observed for $U_{1-x}Th_xBe_{13}$ and $CeRu_2Si_2$. It is only in the former compound that the parabolic character of the μ SR depolarization function at small times has been established. While in UPt_3, magnetic Bragg peaks are seen by neutron and x-ray scattering, most other experimental techniques including μ SR fail to detect a signal of magnetic origin. URu_2Si_2 exhibits a magnetic phase transition with a relatively long correlation length and is characterized by a small uranium magnetic moment. The functional form of $P_s(t)$ confirms that URu_2Si_2 is a relatively well-ordered magnet. The other three widely studied heavy fermion superconductors, UNi_2Al_3, UPd_2Al_3, and $CeCu_2Si_2$, all exhibit relatively large ordered moments and therefore may not belong to the same class of compounds as $CeRu_2$.

$CeRu_2$ appears to be an ordered magnetic superconductor characterized by a small magnetic moment. This invites comparison to similar characteristics in the U-based materials $U_{1-x}Th_xBe_{13}$, UPt_3, and URu_2Si_2. Relative to the latter three compounds, it exhibits even smaller magnetic moments. The shape of the μ SR depolarization function is quadratic in time. While this result does not identify the precise nature of the order, we note that magnetic moments located on the Ru ions would lie on a three-dimensional lattice of corner-sharing tetrahedra: this situation is known to give rise to frustration. This frustration might lead to a glasslike state and would nicely explain the μ SR results. Whatever the nature of the magnetic order, it is likely to influence the pinning of the vortex lattice and presents an important ingredient that needs to be considered to understand the unusual transition from reversible to irreversible behavior in the superconducting state.

Pyrochlore Oxides

The superconductors with the structure of pyrochlore oxide has general formula AOs_2O6, where A= Cs(T_c = 3.3 K), Rb (T_c = 6.3 K) and K (T_c = 9.6 K). However, these compounds with the same chemical formula differ sharply in their superconducting properties. If $RbOs_2O_6$ is the BCS superconductor, then KOs_2O_6 similar to HTSC relates to oxides of transitional metals, but does not crystallize in the perovskite structure. KOs_2O_6 crystallizes in the pyrochlore oxide structure, based on triangle lattice, which is a classic example of frustration effect in spin system, forming numerous spin structures. Generally, pyrochlore oxides present a great group of titan-, tantalum- and niobium-containing minerals with cubic crystals and general formula $A_2B_2O_6O'$, where A is the large cation, B is the smaller cation (usually 5d-transitional metal, i.e., Re, Os or Ir). First, superconductivity in this oxide class has been discovered in $Cd_2Re_2O_7$ compound ($T_c = 1K$). The difference between $Cd_2Re_2O_7$ and KOs_2O_6 is in the number of d-electrons in B-cation. It is interesting that KOs_2O_6 compound with fractional degree of oxidation has the critical temperature 10 times higher than $Cd_2Re_2O_7$, possessing even number of 5d-electrons. In the compounds with the same structure like $RbOs_2O_6$ and $CsOs_2O_6$, Os ion has fractional degree of oxidation, +5.5, disposing in intermediate state between $5d^2$ and $5d^3$. In this case, the 5d-electrons of Os define transport and magnetic properties of these materials, simultaneously, a coupling of which in these oxide compounds is mostly interesting.

The effect of high pressure (up to 10 GPa) on superconductivity in the AOs_2O_6 compounds has been studied for A = Cs, Rb and K. The critical temperature for all three materials increased together with pressure up to a maximum value of $T_c = 7.6$ K (at 6 GPa), $T_c = 8.2$ K (at 2 GPa) and $T_c = 10$ K (at 0.6 GPa) for A = Cs, Rb and K, respectively, after that it diminishes down to total disappearance at 7 and 6 GPa for A = Rb and K, and above 10 for A = Cs.

Rutheno-Cuprates

Sr_2RuO_4 compound ($T_c \sim 1$ K) is the single example of layered perovskite without cooper, demonstrating superconductivity. This compound relates to the class of "self-doped" conductors due to small ratio, U/W (where U is the energy of Coulomb repulsion and W is the width of Brillouin's zone), that is, there the role of electron correlation is not important compared to cuprates. P-type of pairing (spin-triplet) in Sr_2RuO_4 is realized. This system is also called the system with ladder structure. If to seek number of legs of the ladders per cell to infinity, then a transition to two-dimensional structure occurs. In cuprate HTSC, the ladder role could be played by stripes, then, a total analogy between cuprate and ruthenium systems is possible.

Triple perovskite ("hybrid" rutheno-cuprate superconductor) $RuSr_2GdCu_2O_8$ consists of both "superconducting" CuO_2 layers and "ferromagnetic" RuO_2 layers. Herein, superconductivity co-exists with electronic ferromagnetism in microscopic scale (the temperature of ferromagnetic ordering 135 K and $T_c = 50$ K). The investigations of the magnetization and magnetic resistance of the $RuSr_2GdCu_2O_8$ have demonstrated influence of the magnetic moments of Ru atoms on the electrons of conductivity.

The intragrain critical temperature, T_c, of HTCS cuprate $RuSr_2(Gd, Ce)_2Cu_2O_{10+d}$ has been studied depending on hole concentration (which is found by oxygen content). In this case, T_c changed in very wide limits (17–40 K) with the change of p being only 0.03 holes/CuO_2. Into this range of p, the intragrain superfluid density (which is inversely proportional to square of the magnetic field penetration depth, $1/\lambda^2$) and the value of the diamagnetic jump (found at the sample cooling in magnetic field) have increased more than 10 times. These results contradict to correlations between T_c, p and $1/\lambda^2$, observed in homogeneous HTSC. This is possible due to the phase de-lamination and granularity. Moreover, there is an effect of anomalous increasing of distance between CuO_2 planes during cooling of layered compound $RuSr_2Nd_{0.9}Y_{0.2}Ca_{0.9}Cu_2O_{10}$, doped up to level of the boundary "antiferromagnetism/superconductivity" at the phase diagram. This means negative value of the thermal expansion factor. Difference in volumes of the antiferromagnetic and super-conducting phases may be caused by the phase segregation, observed often in weakly doped HTSC.

Magnetic Superconductors

In 1971, Chevrel and co-workers discovered a new class of ternary molybdenum sulfides having the general chemical formula $M_xMo_6S_8$, where M stands for a large number of metals and rare earths. These superconductors have unusually high values of the upper critical field, B_{c2}, given in table. The superconducting compounds $REMo_6X_8$ (where RE = Gd, Tb, Dy, Er and X = S, Se, Te), and $RERh_4B_4$ (where RE = Nd, Sm, Tm) are usually related to the Chevrel phases. Magnetic super-conductors demonstrate some novel features not found in conventional type-I superconductors. Upon cooling from the superconducting phase ($T < T_c$), the material becomes normal again at a low temperature and the superconductivity is destroyed. Very often this normal phase is magnetically ordered. Upon cooling from the normal state the system becomes superconducting below T_c, and upon further cooling it becomes magnetically ordered below Neel temperature, T_N (where $T_N < T_c$). Thus, the superconducting phase occurs only in a limited range of temperatures, $T_N < T < T_c$. The interaction of conduction electrons with magnetic atoms leads to the formation of a bound state of the electron with the magnetic atom below a certain characteristic temperature called the Kondo temperature. It was found that the resistivity of a magnetic alloy, such as Fe impurities in Cu shows a minimum in the resistivity as a function of temperature due to the interaction of conduction electrons of the metal atom with the magnetic moment of the magnetic atom. Such a Kondo effect exists in superconductors containing magnetic impurity atoms at low temperatures, because the superconductivity is destroyed. In the case of magnetic superconductors, there is a sub-lattice of magnetic atoms in addition to the lattice of metallic atoms, so that magnetically ordered phase exists at low temperatures below the superconducting phase, $T_N < T_c$.

Table: The critical temperature and the upper critical magnetic field of Chevrel phases.

No.	Compound	$T_c(K)$	$B_{c2}(T)$
1	$PbMo_6S_8$	15	60
2	$LaMo_6S_8$	7	44.5
3	$SnMo_6S_8$	12	36

While ferromagnetic superconductors (e.g., $ErRh_4B_4$ and $HoMo_6S_8$) demonstrate the above be-havior, the antiferromagnetic superconductors (e.g., $REMo_6S_8$, where RE = Gd, Tb, Dy and Er and also $RERh_4B_4$, where RE = Nd, Sm and Tm) present examples of the co-existence of the two phases together with anomalous behavior of the second critical field. In $ErRh_4B_4$, possessing tetragonal structure with a = 5.299 °A and c = 7.588 °A, a plot of the ac-magnetic susceptibility (χ_{ac}) and elec-tric resistance depicts a normal to superconducting transition at T_{c1} = 8.7 K, followed by a loss of susceptibility at T_{c2} = 0.9 K together with the appearance of the ferromagnetic long-range order. This result takes place at zero magnetic field. At non-zero finite field also, the resistance disap-pears at some temperature interval within the above domain. The compound HoMo6S8 becomes superconducting at T_{c1} = 1.3 K in zero field. On further cooling, it becomes normal at T_{c2} = 0.6 K together with the appearance of ferromagnetic long-range order.

The antiferromagnetic superconductors provide the most striking case of the co-existence of the two kinds of order. These systems (e.g., $REMo_6S_8$, where RE = Gd, Tb, Dy, Er and also $RERh_4B_4$, where RE = Nd, Sm, Tm) demonstrate near TN the antiferromagnetic alignment of rare-earth magnetic moments in the superconducting state of the system. The most important result is the anomalous behavior of the upper critical field as a function of temperature near TN. In particular, (RE = Gd, Tb and Dy) demonstrate anomalous decreasing of H_{c2} near, but below TN. The crystal structure of $REMo_6S_8$ is presented in figure.

An attempt of T_c increasing in REMo6S8 owing to change of RE ion radius leads to the structure transition, as a result of which the dielectric gap opens at Fermi level and the superconducting properties disappear together with metallic ones. Based on BCS model, this is explained that the growth of T_c is caused by the approaching Fermi level to the peak of the electron state density, N(E). However, it is disadvantageous energetically for the structure, therefore it suffers phase transition, which suppresses superconductivity.

The superconducting carbosulfide Nb_2SC_x with layered crystalline lattice has critical temperature $T_c = 5$ K at x = 0.8–1. The superconductivity demonstrates volume character, and Nb atom octahedron is the key structure element, in the center of which C atom locates.

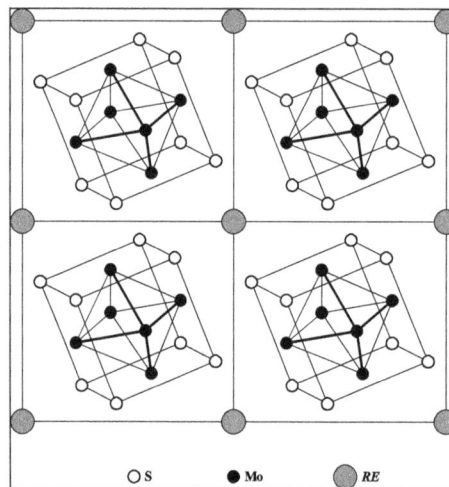

Structure of $REMo_6S_8$, which have Mo octahedron inside sulphur cube, which are inside a rare-earth cube.

Heavy Fermion Superconductors

Heavy fermion materials are a class of compounds named for the enormous effective mass of their charge carriers.

The Kondo Lattice Model

In order to understand heavy fermion superconductors, the Kondo lattice model is a good starting point. In order to understand the behavior of magnetic impurities in metals, the Anderson Hamiltonian describes the coupling of electrons in the outer orbital of the magnetic impurity to the conduction electrons of the metal.

The Anderson Hamiltonian comes in three parts. The first part is simply the kinetic energy of the conduction electrons. The second part is the Hamiltonian of electrons on the atom - both a binding energy and an on-site repulsion. The third part is a magnetic interaction term between the electrons on the atom and the conduction electrons. Left to themselves, the atoms will form a magnetic ground state. But with the introduction of the interaction term, that's no longer an energy eigenstate - the atom slowly exchanges spins with the surrounding conduction electrons.

At low energies, related to the frequency ω of this spin exchange, there is a resonance which leads to an increase in the scattering cross-section, and thus an increase in the resistance. The Kondo temperature T_K is the characteristic temperature at which this effect becomes large, determined by $k_B T_K = \hbar\omega$.

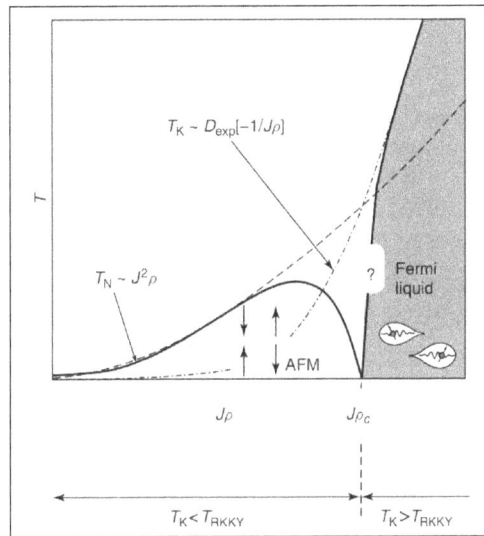

Diagram of T_K and T_{RKKY} (T_N). When $T_K > T_{RKKY}$, the antiferromagnetic phase is replaced by a heavy fermion liquid.

After the discovery of heavy fermion metals in 1975, Doniach proposed that their properties were the result of a dense lattice of magnetic moments, all screened by the Kondo effect. A model of a lattice magnetic moments interacting with the conduction electrons had previously been worked out in the 1950s. The magnetic moments induce waves in the electron spin, which can then interact with other magnetic moments. This RKKY interaction, named after Ruderman, Kittel, Kasuya, and Yosida, usually leads to an antiferromagnetic ordering characterized by the energy $k_B T_{RKKY}$. Doniach proposed that when $T_K > T_{RKKY}$, which can occur at high coupling strength and carrier density, the Kondo effect could significantly change the properties. At low temperatures the magnetic moments would be a lattice of resonant scattering centers, with scattered electrons phase-shifted by a constant. Bloch wave functions would still be possible, but they would have to incorporate this resonant scattering, leading to a band of roughly the same width as the resonance, or $\sim k_B T_K$. This narrow band would have to have high curvature, leading to a high effective mass.

Measured Properties

There are many heavy fermion superconductors to choose from, with a diversity of properties that is very large compared to the cuprates. Different heavy fermion superconductors may even be best described by different models of superconductivity. Still, there does seem to be a strong connection between heavy fermion materials and superconductivity, and there are a few general properties shared by heavy fermion superconductors.

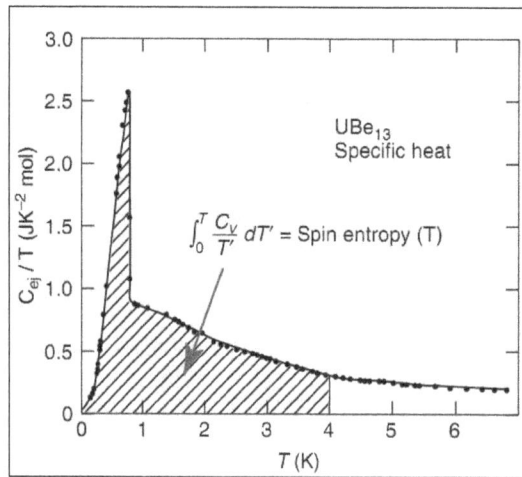

The specific heat of UBe$_{13}$ as it goes through its superconducting transition. If the superconducting state did not have heavy fermion properties, the specific heat below the jump would be much smaller.

One general property of heavy fermion superconductors is that the superconducting electrons seem to be the same ones involved in the Kondo effect. Clear evidence for this is given by specific heat capacity measurements, because heat capacity is scaled by the effective mass of the quasi-particles. The specific heat capacity of heavy fermion superconductors jumps at T_c, but the scale remains set by the large effective mass of the quasiparticles.

CeCu$_2$Si$_2$ was the first confidently observed heavy fermion superconductor. It has a Kondo temperature of approximately 10 K, and becomes superconducting at $T_c = 0.7$ K. Before the superconducting transition, it enters into an unusual "A phase" which has slow magnetic fluctuations, as measured by nuclear magnetic resonance and muon spin experiments. There is no nearby magnetically ordered phase, but the similarities under pressure to CeCu$_2$Ge$_2$ suggest that the material is just past being antiferromagnetic. The strong reduction in magnetic susceptibility below T_c signals a singlet pairing state. The symmetry of the order parameter is a bit unclear, with most low-temperature thermodynamic properties behaving as if the order parameter vanished along line nodes on the Fermi surface.

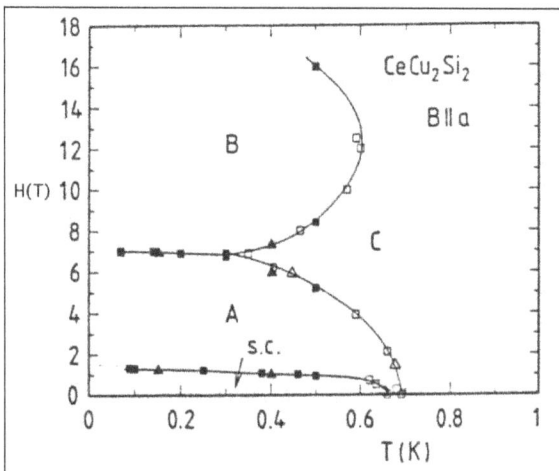

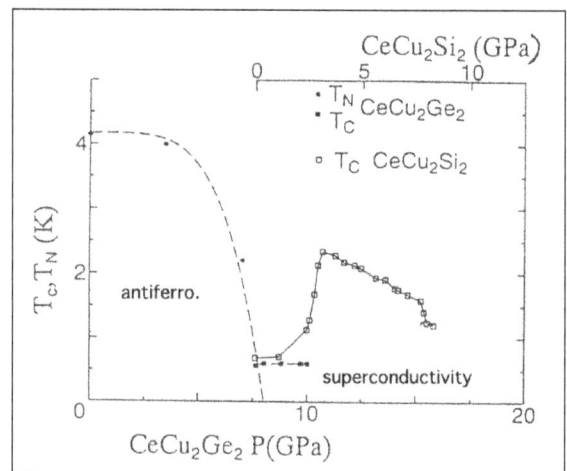

Figure shows the graph on the left shows the phase diagram of CeCu$_2$Si$_2$ under magnetic field. Note the ordered A phase that surrounds the superconducting phase. The graph on the right shows a comparison between $CeCu_2Si_2$ and $CeCe_2Ge_2$ under pressure – the rough similarity suggests that $CeCu_2Si_2$ starts out tuned past antiferromagnetic order.

UBe_{13} has a cubic crystal structure dominated by huge uranium atoms, with a T_C of 0.86 K. Like $CeCu_2Si_2$, and like another uranium-containing heavy fermion superconductor UPt_3, UBe_{13} is not near any magnetically ordered states, and unlike in $CeCu_2Si_2$ there is no additional reason to think that magnetism is close by even in an inaccessible way. Because of an inconclusive shift in magnetic susceptibility and a very high upper critical field, the pairing state is likely spin-triplet. Specific heat and NMR measurements disagree about the type of nodes in the order parameter, suggesting line and point nodes respectively. When doped with thorium, the T_C of UBe_{13} decreases up to a point, then increases temporarily when it enters an unusual region where there are actually two phase transitions. The normal state of UBe_{13} is unusual, with some non-Fermi liquid properties predicted by a high-entropy "incoherent metal" model.

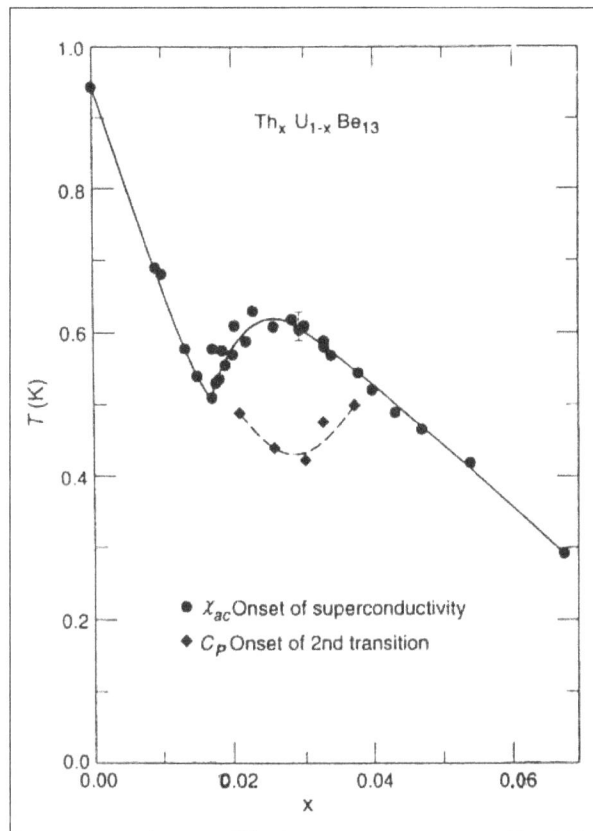

the unusual behavior of UBe_{13} under doping with thorium. When $0.19 < x < 0.4$, UBe_{13} has a poorly understood second phase transition.

$CeIn_3$ was the first discovered "quantum critical" heavy fermion superconductor. Normally it forms antiferromagnetic order at temperatures below 10 K, but as pressure is increased the antiferromagnetic transition temperature goes to zero. Near the projected quantum critical point, the properties of the normal state no longer scale with temperature as predicted by the Fermi liquid model, and indication of the quantum critical region. Superconducting order in $CeIn_3$ forms a dome centered on the quantum critical point, with $T_c = 0.2$ K at 2.5 GPa. The pairing state is singlet, and the order parameter has line nodes on the Fermi surface.

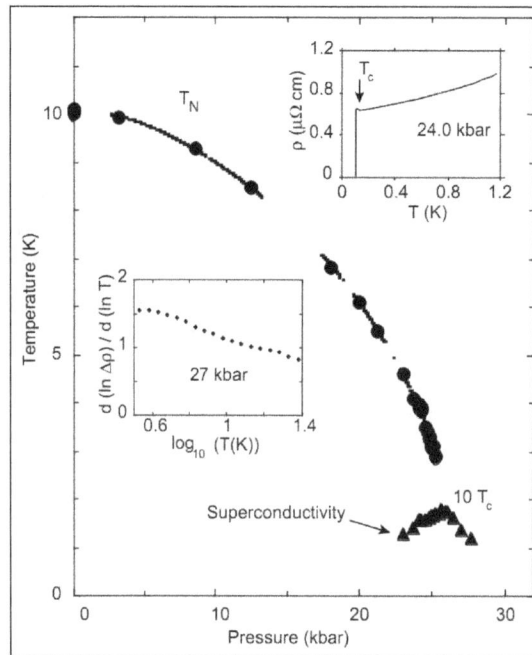

The phase diagram of CeIn$_3$ under pressure. As the antiferromagnetic transition temperature
TN goes to zero near 25 kbar (2.5 GPa), a superconducting phase appears.

CeCoIn$_5$ was the first of the "1-1-5" class of heavy fermion materials. It has a layered structure, with conduction occurring in layers that are similar to CeIn$_3$. This type of structure had been predicted to enhance T$_c$, and so it did, to 2.3 K under no pressure and 2.6 K under 1.3 GPa. The phase diagram actually starts to resemble that of the cuprates, with a pseudogap regime appearing below optimal pressure. CeCoIn$_5$ exhibits no magnetic order, but like for CeCu$_2$Si$_2$ there are speculations about antiferromagnetic order at "negative pressure," based in analogy to the related material CeRhIn$_5$. The normal state of CeCoIn$_5$ is non-Fermi liquid, similar to CeIn$_3$. The pairing state is singlet, and the symmetry of the order parameter is probably d$_{x2-y2}$, in the superconducting planes.

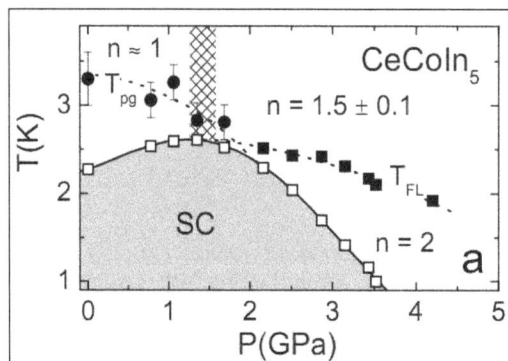

The phase diagram of CeCoIn$_5$ under pressure. Below T$_{pg}$, the system seems to be in a pseudogap state analogous to the cuprates. Below T$_{FL}$, the system is in a heavy Fermi liquid state.
Above T$_{FL}$ and T$_{pg}$, the system is in a non-Fermi liquid normal state.

Another notable 1-1-5 material is PuCoGa$_5$, which has the very high T$_c$ of 18.5 K. Possibly because T$_c$ is roughly as large as the Kondo temperature, PuCoGa$_5$ makes a direct transition from normal metal to heavy fermion superconductor. No evidence for a nearby magnetically ordered state was found. Similar to CeCoIn$_5$ it forms singlet pairs, and has line nodes in the order parameter.

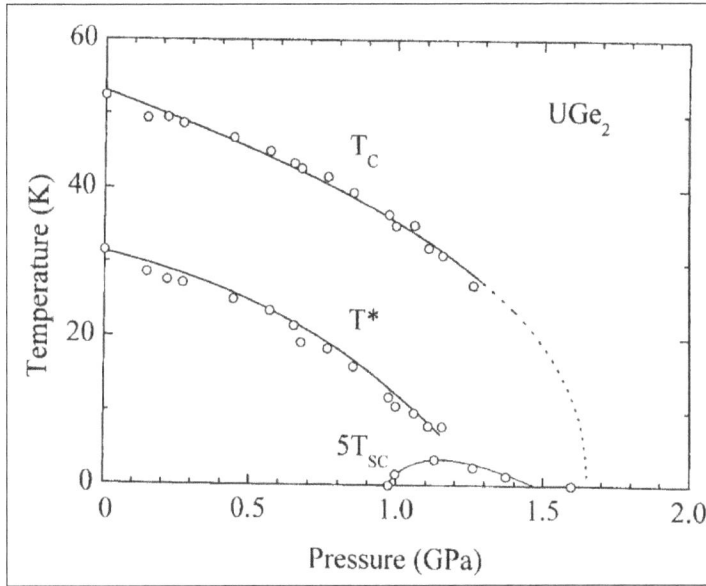

The phase diagram of UGe$_2$ under pressure. TC is the ferromagnetic transition temperature, while T* is an transition in electronic order. The superconducting dome appears around the quantum critical point of T*, within the ferromagnetic region.

UGe$_2$ is interesting because it has ferromagnetic order rather than the usual antiferromagnetic order. It reaches a maximum T$_c$ of 0.7 K at 1.1 GPa. UGe$_2$ appears to be a quantum critical superconductor, but the quantum critical point is not the disappearance of ferromagnetic order. Instead, there's some sort of electronic order that appears to go to zero at the maximum T$_c$. The heavy fermion superconducting state appears to coexist with ferromagnetism, making a triplet pairing state likely. Antiferromagnetic ordering has also been observed to coexist with the superconducting state, in the heavy fermion superconductor URu$_2$Si$_s$.

U-based Superconducting Compounds

The first two U-based heavy-fermion superconductors, UBe$_{13}$ (T$_c$ = 0.9° K) and UPt$_3$ (T$_c$ = 0.54° K), were discovered in 1983 by Ott et al. and in 1984 by Stewart et al. respectively. It was evident within a few years that UPt$_3$ had three superconducting phases, which created great impetus for further study of this unusual heavy-fermion superconductor.

UBe13 was the first actinide-based heavy-fermion compound that was found to be a bulk superconductor below approximately 0.9° K. The cubic UBe$_{13}$ is also one of the most fascinating HF superconductors because superconductivity develops out of a highly unusual normal state characterized by a large and strongly T-dependent resistivity. In addition, upon substituting a small amount of Th for U in U$_{1-x}$Th$_x$Be$_{13}$, a nonmonotonic evolution of T$_c$ and a second-phase transition of T$_{c2}$ below T$_{c1}$, the superconducting one, is observed in a critical concentration range of x.

It was also shown that the superconducting state is formed by heavy-mass quasiparticles. This was demonstrated by plotting C$_p$/T versus T (at low temperatures), which is shown in figure. The anomaly at T$_c$ is compatible with the large γ parameter in the normal state at this temperature.

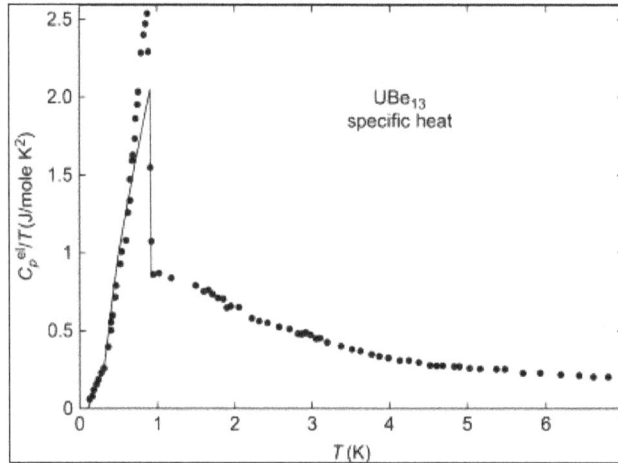

Electronic specific heat of UBe$_{13}$ below 7°K. The solid line represents
the BCS approximation of the anomaly at and below T$_c$.

The temperature dependence of the specific heat of UBe$_{13}$ well below T$_c$ was the first indication of the unconventional superconductivity. Figure shows the non-exponential but power-law-type decrease of $C_p(T)$ that was interpreted as being the consequence of nodes in the gap of the electronic excitation spectrum.

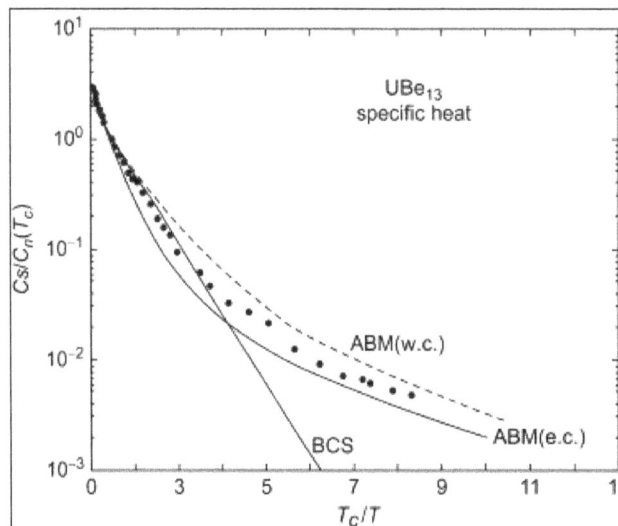

Normalized electronic specific heat of UBe$_{13}$ below T$_c$, plotted versus T$_c$/T. The solid
and broken lines represent calculations assuming point nodes in the gap.

It was also found that when small amounts of U atoms in UBe$_{13}$ were replaced with other elements, there was a substantial reduction of the critical temperature. T$_c$ is also first substantially reduced with the alloys U$_{1-x}$Th$_x$Be$_{13}$ as x is increased. However, when x > 0.018, T$_c$ increases again until it passes over a willow maximum at x = 0.033 and gradually decreases with a reduced slope when x is further increased. Further, in the range 0.019 < x < 0.05, a second transition at T$_{c2}$ below T$_c$ was discovered by measuring the specific heat of these alloys at very low temperatures. Measurements of ρ(T) and χ(T) confirmed that the phase at temperatures below the second anomaly of C$_p$(T) was superconducting.

The phase diagram of superconductivity of U$_{1-x}$Th$_x$Be$_{13}$, from these observations as well as from thermodynamic arguments. One can identify three different superconducting phases: F, L, and U.

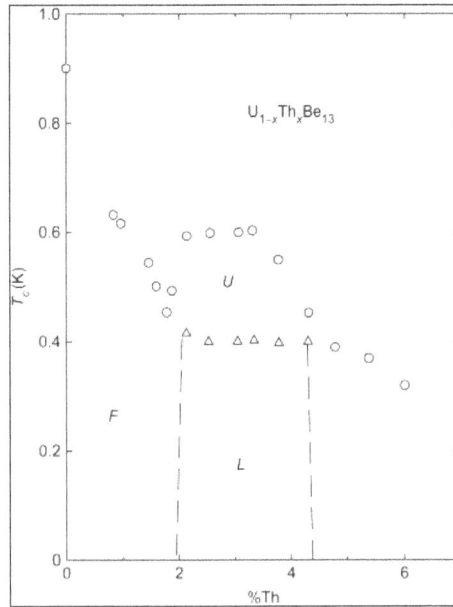

An x, T phase diagram for superconducting $U_{1-x}ThxBe_{13}$ as derived from the measurements of the specific heat. The letters F, L, and U denote three superconducting phases.

Superconductivity has also been discovered in UPt_3, and its alloys, URu_2Si_2; and in UPd_2Al_3 and its alloys, UNi_2Al_3, UGe_2, URhGe, and UIr.

The discovery of superconductivity in UGe_2 in single crystals of UGe_2 under pressure below $P_c \sim 16$ kbar was very surprising. The sensational part of this discovery is that the pressure $p \sim 12$ kbar, where the superconducting temperature TS $= 0.75°$ K is strongest, the Curie temperature $T_C \sim 35°$ K is two orders of magnitude higher than T_S; superconductivity occurs in a very highly polarized state ($\mu(T \rightarrow 0°K) \sim \mu_B$).

The superconductivity in UGe_2 disappears above a pressure $P_c \approx 16$ kbar that coincides with the pressure at which the ferromagnetism is suppressed. The pressure-temperature phase diagram of UGe_2 is shown in figure.

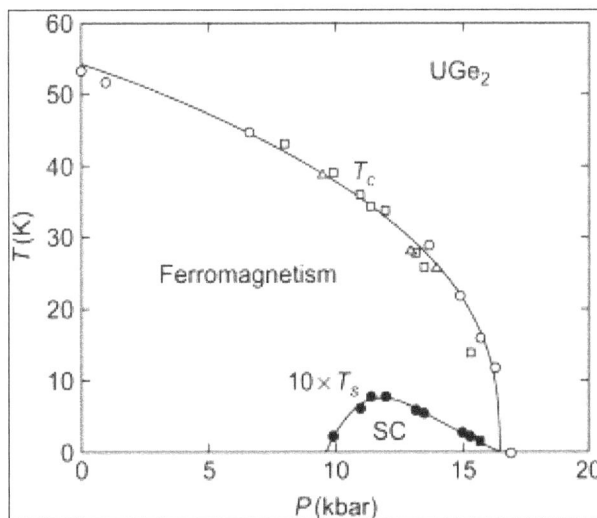

The pressure-temperature phase diagram of UGe_2.

Oxide Superconductors without Copper

The superconductors of A_xWO_3 type have hexagonal structure of the tungsten bronze, whereas A, the large alkaline ions of K, Rb and Cs are used more often. Many superconducting materials from this family demonstrate $T_c \approx 2$–7 K. The monocrystals of perovskite dielectric WO_3, doped by Na, demonstrate in surface layer of $Na_{0.05}WO_3$ high-temperature superconductivity with $T_c = 91$ K. In this case, Na segregates at the grain surface and takes the structure of the WO_3 surface layer.

The superconductor without copper, $Ba_{1-x}K_xBiO_3$ (BKBO) demonstrates maximum critical temperature ($T_c = 31$ K) among "old" oxide superconductors. The oxide superconductor, $LiTi_2O_4$, without Bi and Cu, also possessing high transition temperature, has the spinel crystalline structure. Two other oxide superconductors: NbO and TiO, showing $T_c \leq 1$ K, demonstrate clearly the expanded links of the "metal-metal" type as the previous oxide. The oxide superconductor $BaPb_{0.75}Bi_{0.25}O_3$ with the critical temperature $T_c = 12$ K has simple perovskite structure. Recently, the layered superconductor $Na_xCoO_2 \cdot 1.3H_2O$ with $T_c = 4$ K has been fabricated using the chemical oxidation technique. The conductive layers (CoO_2) in this superconductor alternate with the dielectric buffer layers ($Na_x \cdot 1.3H_2O$), which carry out a function of reservoirs for electric charge. Similar to HTSC, the value of T_c is highest at the optimum level of doping and decreases in overdoped and underdoped samples.

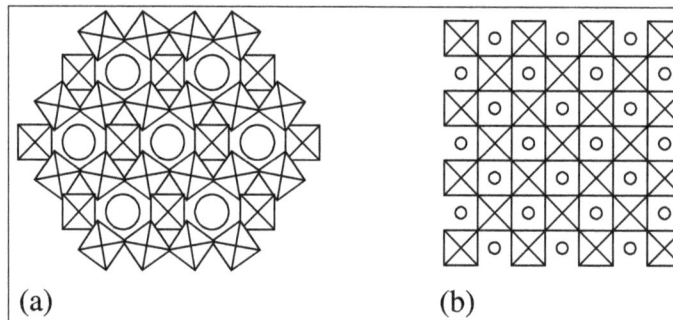

(a)　　　　　　　　　　　　　　　　　(b)

Hexagonal structure of the tungsten bronze. One elementary layer presented, which is formed by the octahedrons MO_6 with large ions in cavities. (b) Perovskite structure.

Rare-Earth Borocarbides

The rare-earth borocarbide $LuNi_2B_2C$ has critical temperature, T_c, approximately 16 K, at the same time, LuNiBC is not superconducting. The superconductor is obtained, adding carbon in Lu plane. On the other hand, LuNiBC is obtained from $LuNi_2B_2C$, adding the other layer of Lu–C. Thus, the superconducting composition is fabricated from isolator by changing corresponding atom ratio. Similarly in $La_{2-x}Sr_xCuO_4$, superconducting phase is formed from isolating phase by changing parameter x. The compounds $ReNi_2B_2C$ with Re = Y, Lu, Tm, Er, Ho and Dy are superconducting with moderate high critical temperatures, $T_c \approx 16$ K. At low temperatures, there is antiferromagnetic phase, and the phase exists in which can co-exist superconductivity and magnetic long-range order. As for the example of $Ho_{1-x}RE_xNi_2B_2C$ (RE = Y, Lu) system with different numbers of Ho(RE)C layers, a clear correlation of T_c with the density of states in Fermi level has been found. The various magnetic structures are derived which co-exist with superconductivity. The highest T_c = 23 K is attained in YPd_2B_2C.

Crystal Structure and Critical Temperature

The tetragonal layered crystal structures of the I4/mmm or P4/nmm types resolved so far for all well characterized RTBC(N) compounds can be written schematically as $(RC(N))n(TB)_2$ with $n = 1,2,3$. There are systematic dependences of T_c with increasing T–T distance, the transition metal component T = Ni, Pd, Pt and the dopants replacing the T: T' = Cu, Co, V, etc., and the B–T–B bond angle. Finally, the number of metallic layers separating and doping the $(NiB)_2$ networks also has a profound effect on the actual T_c value. Thus, for the single c RC-layer (T = Ni) compounds the highest $T_c \approx 14 - 16.6\,K$ values are obtained for R = Sc, Y, Lu, whereas for R = Th it is reduced to 8 K and it vanishes for R = La. The double-layer Lu, Y-compounds exhibit very small transition temperatures of 2.9 K and 0.7 K, respectively, which however can be increased considerably replacing Ni by Cu. In the case of the two-layer boronitride $(LaN)_2(NiB)_2$ so far no superconductivity has been detected whereas the corresponding triple-layer compound exhibits a relatively high Details of what appear to be the cubic crystal structures of several metastable Pd-based compounds showing in some cases $T_c \geq 21\,K$ have not been resolved till now.

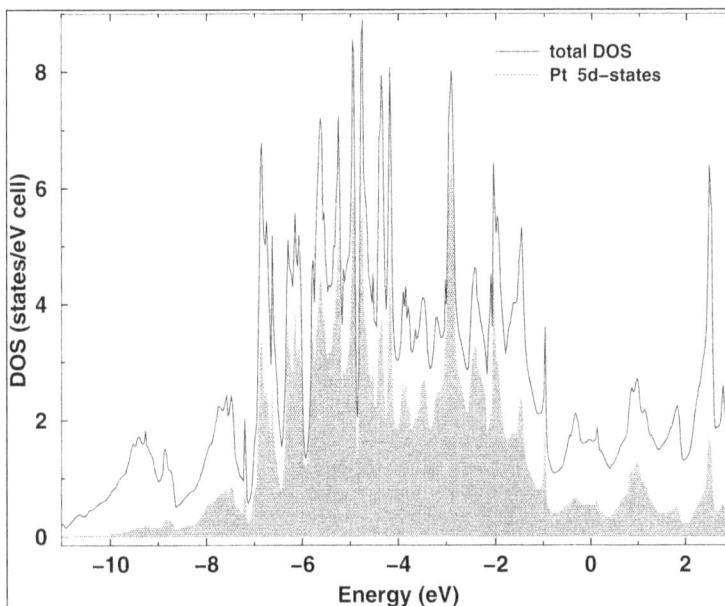

LCAO–LDA full relativistic total DOS (solid line) and Pt-partial 5d DOS (shaded) of $LaPt_2B_2C$; with the Fermi level EF = 0.

The Electronic Structure

All bandstructure calculations reveal sizeable dispersion in c-direction of the bands crossing the Fermi level and fluctuation magnetoconductance measurements clearly demonstrate the 3D nature of the superconductivity under consideration. Electronically the coupling of the 2D-$(TB)_2$ networks is mediated mainly by the carbon/nitrogen 2 pZ states. Further, important issues are the peak of the density of states (DOS) N(0) near the Fermi level EF = 0 and the intermediate strength of correlation effects.

LDA Calculations

The electronic structure near EF = 0 of all RTBC(N) compounds is characterized by a special band complex containing three or four bands (total width about 3 eV for the case of LaPt$_2$B$_2$C above the main group of T-derived d-states (total width about 7 eV). For Lu(Y)Ni$_2$B$_2$C there is a flat band near EF giving rise to a narrow asymmetric peak in the total DOS N(E) and to a large T-d-partial DOS $N_d(0) \approx 0.5$ N(0) which can be analyzed with high accuracy within our LCAO-method (linear combination of atomic-like orbitals) calculational scheme. For most of the other RTBC-superconductors N(0) and especially Nd(0) are reduced. However, there is no simple relation between the calculated value of N(0) and the measured T$_c$-value. Notice that our full relativistic calculation for LaPt$_2$B$_2$C yields a significantly reduced DOS $N(0) = 1.66$ states/eV cell compared with 2.5 obtained by Singh where only the core states were treated fully relativistically and the valence states scalar relativistically. For our N(0) the previous discrepancy with the specific heat data for LaPt$_{1.5}$Au$_{0.5}$B$_2$C and LaPt$_{1.7}$Au$_{0.3}$B$_2$C for the Sommerfeld constant $\gamma \alpha N(0)(1+\lambda)$ is removed: instead of an unreasonably small electron–phonon (el–ph) coupling constant $\lambda \approx 0.1$, we arrive now at more realistic values $\lambda \approx 0.66$ to 0.94, respectively. Our calculations for Lu(Co$_x$Ni$_{1-x}$)$_2$B$_2$C within the LCAO-coherent potential approximation show a strong reduction of N(0) by about 21% and 42% for x = 0.1 and x = 0.5, respectively. The decrease of the Ni + Co related partial d-DOS is even stronger. All in all, the comparison with the available specific heat data reveals that most RTBC(N)-compounds exhibit intermediately strong el–ph interaction $\lambda \sim 0.5$ to 1.2, except LaPd$_2$B$_2$C exhibiting extremely weak el–ph coupling.

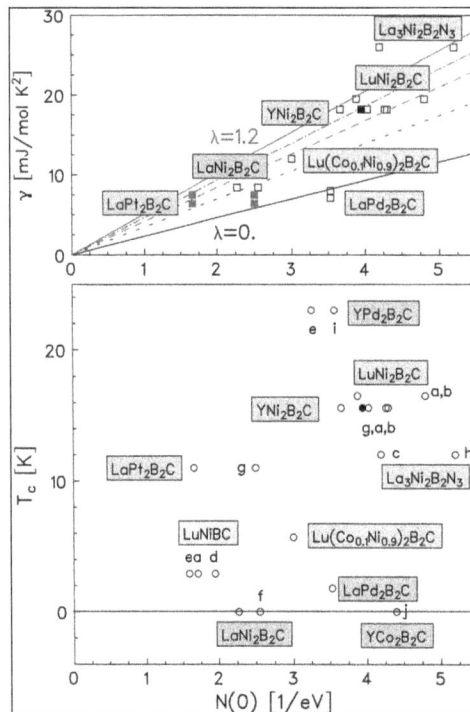

Experimental values of the Sommerfeld constant (top) and the superconducting transition temperature T$_c$ (bottom) vs. calculated total density of states at the Fermi level N(0) (states per formula unit) for various RTBC(N)-compounds and different LDA-calculational schemes: The straight lines in the upper picture denote various el–ph coupling constants $0 \le \lambda \le 1.2$ (dotted line: λ = 0.5, dashed line: λ = 0.8, dashed–dotted line: λ = 1.0.

Electronic Spectroscopy and de Haas–van Alphen Effect

Polarization dependent X-ray absorption spectroscopy (XAS) for an YNi_2B_2C single crystal probing the unoccupied electronic structure via transitions from the B(C) 1s core level into unoccupied states having B(C) 2px/2pz symmetry are in reasonable agreement with the predictions of the LDA bandstructure calculations for the orbital-resolved partial DOS. The covalency of the short C–B bond seems to be somewhat overestimated by the LDA predicting a smaller out-of-plane anisotropy than is observed in the XAS data. The observed isotropic Ni-related spectra are presumably dominated by effects of the strong 2p core hole 3d Coulomb interaction. Therefore no direct comparison with the LDA prediction is possible. From a comparison of X-ray photoemission and Auger spectroscopy measurements a value of the Ni d–d Coulomb repulsion $U_d = 4.4\pm0.3\,eV$ has been found which is w x d intermediate between the values of 8 eV and 2 eV for strongly correlated NiO and weakly correlated Ni metal, respectively.

The most valuable insight into the electronic structure can be gained from de Haas–van Alphen (dHvA) measurements. In high-quality YNi_2B_2C single crystals up to six cross-sections are observed. The related Fermi velocities F,i, i = 1, . . . ,6, on extremal orbits can be grouped into two sets differing by a factor of 4. These observations and the sizable anisotropy of the H_{c2} for such crystals clearly indicate that they are nearly in the clean-limit regime. Further details of the Fermi surface can be obtained from the analysis of Hall data. However, the observed qualitative differences between YNi_2B_2C and $LuNi_2B_2C$ as well as the large discrepancy between the experimentally and theoretically predicted RH-values are not well understood at present. To summarize, at the present status, there is a reasonable qualitative agreement between predictions of the LDA results and experimental data.

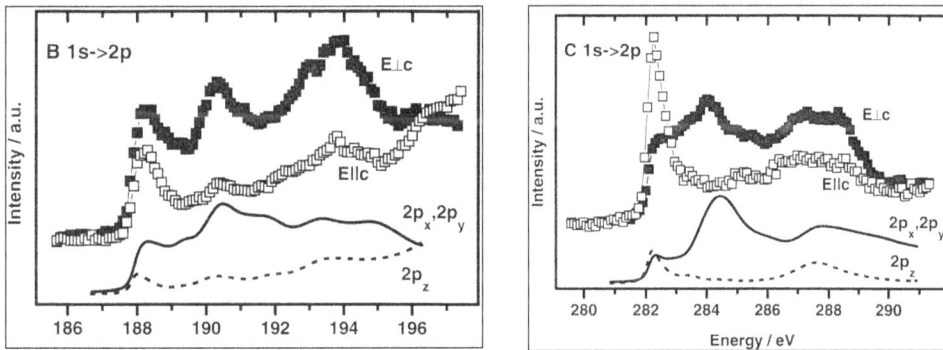

Polarisation-dependent XAS spectra ($1s \rightarrow 2p$ transitions) of single crystal YNi_2B_2C are shown for boron (top) and carbon bottom with the electric field and $\vec{E}\,\|$ to the tetragonal c-axis (open and filled symbols, respectively). The corresponding m-resolved partial DOS from our LDA–LCAO calculations are denoted by dashed and full lines, and are broadened by account for life time and finite resolution effects.

The Upper Critical Field $H_{c2}(T)$

The failure of the standard isotropic band approach points to a multiband description, where electrons with significantly smaller compared with the Fermi surface average $\sqrt{\langle v_F^2 \rangle}$FS and relatively strong el–ph coupling are mainly responsible for the superconductivity. This model explains also the strong deviations of the shape of $H_{c2}(T)$ from the standard parabolic-like curve and in particular also the positive curvature of $H_{c2}(T)$ near T_c. This curvature has very often been observed in resistivity measurements. In principle, it might be affected also by the flux state in fields $H \leq H_{c2}$. However, the fact that it has also been observed in magnetization and specific heat measurements shows unambiguously that it is an inherent thermodynamic property generic for all clean RTBC(N) superconductors.

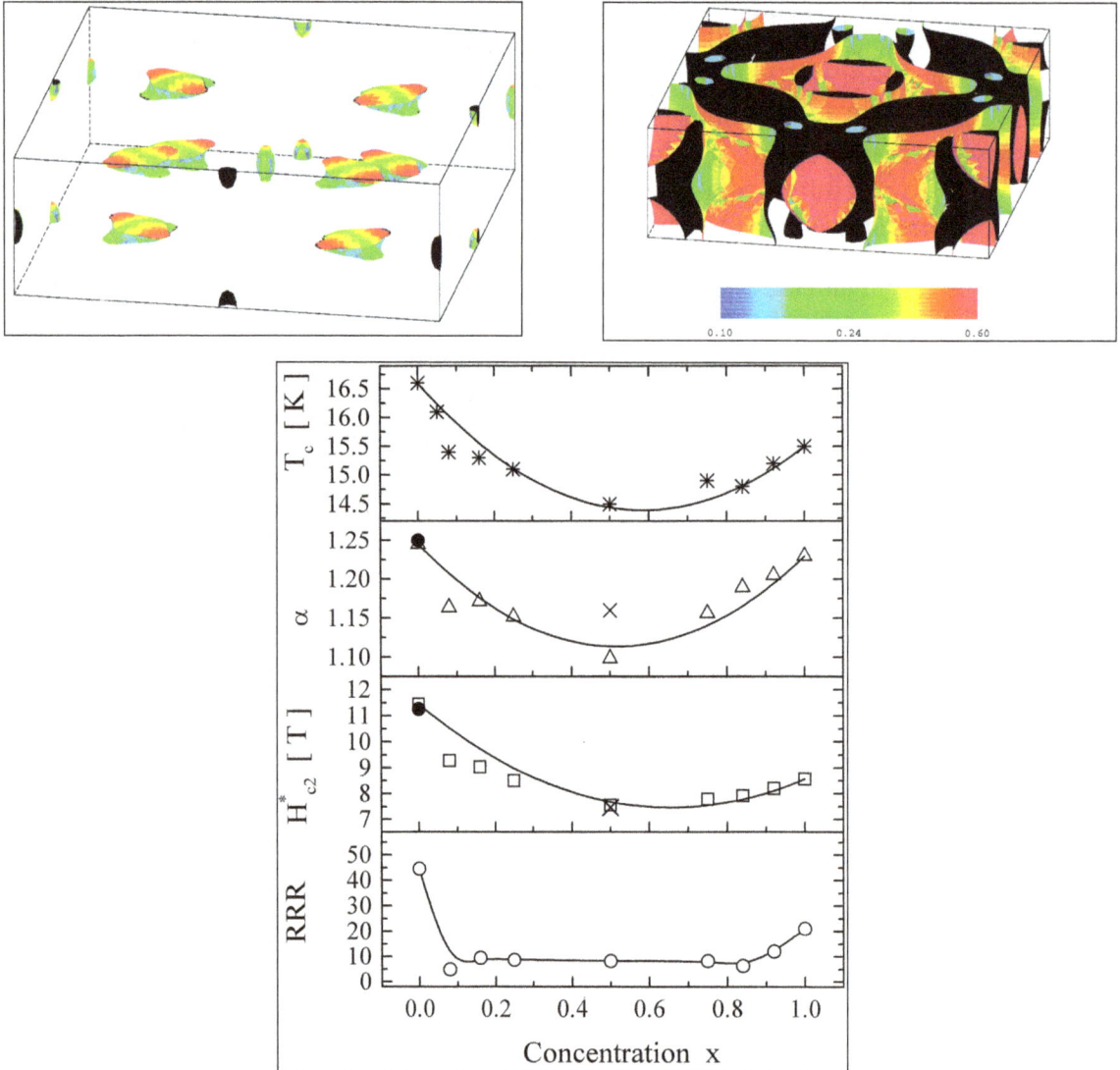

The superconducting characteristics: T_c, the upper critical field parameters $H_{c2}^{*}(0)$ and the curvature parameter determined from resistivity measurements and the resistivity ratio RRR for the mixed system $Y_x Lu_{1-x} Ni_2 B_2 C$ vs. composition x. $\times (\bullet)$ represents the results derived from the inset, from the specific heat jump at and from the magnetization.

The inspection of the distribution of v_F over the F Fermi surface of $LuNi_2B_2C$ derived from our LCAO–LDA calculations depicted in figure yields that the major part of the electrons (dominating the normal state transport properties) exhibit large -values (red and green) differing up to factor of 6 from the slow electrons (blue). This is in accord with our phenomenological analysis of $H_{c2}(T)$ requiring a factor of 5 in their relative magnitudes. The electrons with low v_F's are found near those parts of the Fermi surface with nesting properties. Quite interestingly, vectors closely related to the nesting vector $q \approx 0.62\pi / a$, i.e., connecting the neighbouring blue parts, seem to occur also in low-frequency phonons exhibiting anomalously strong softening entering the superconducting state as well as in the incommensurate a-xis modulated magnetic structure which partially suppresses the superconductivity in low magnetic fields.

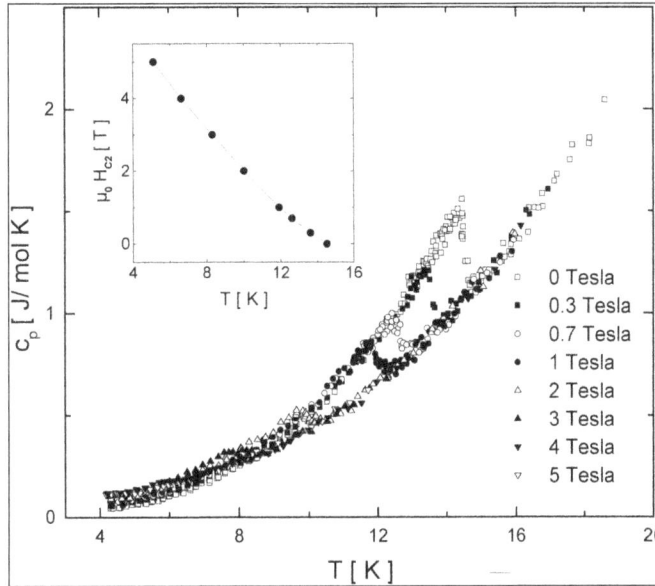

The specific heat of $Y_{0.5}Lu_{0.5}Ni_2B_2C$ vs. temperature T. The inset shows the upper critical field determined from the jump at $T_c(H)$.

Disorder and Doping

The reduction of the multiband effects discussed above can be studied by replacing partially some of the constituent atoms by entities with similar chemical and physical properties. Modest effects can be expected for chemically and magnetically equivalent substitutions R → R' in the RC(N)-layer(s). For this purpose we have studied the crystal structure, T_c, $H_{c2}(T)$, and the specific heat of the mixed poly-c_2 crystalline $YxLu_{1-x}Ni_2B_2C$ system. Since the measured lattice constants a and c vary nearly linearly between the corresponding pure limits, a maximum of Tc might be expected for x = 0.5 according to the 'universal' curve $T_c = T_c(d_{Ni-Ni})$, where d_{Ni-Ni} denotes the Ni–Ni distance.

Instead of the expected maximum with Tc ≈ K we found a dip at x = 0.5 with $T_c \approx 14.5$. Also a reduction of the positive curvature which can be expressed conveniently over a wide temperature range by the exponent α

$$H_{c2}(T) \approx H_{c2}^*(0)(1-T/T_c)^\alpha,$$
$$0.3 \leq T_c \leq 0.95 T_c.$$

on going from high quality pure samples to mixed ones: $\alpha \approx 14.5$ (single crystals), 1.25 (polycrystalline samples) to $\alpha_p \approx 1.1$, $H_{c2}^*(0) = 7.46$ T derived from U resistivity data and $\alpha_c \approx 1.16$, $H_{c2,C}^*(0) = 7.46$ from and specific heat data is worth mentioning. Due to the saturation (negative curvature) at low T, $H_{c2}^*(0)$ can be regarded as an upper bound for the true (0) -value.

The residual resistivity ratio $RRR = \rho(300\,K)/\rho(T_c)$ is strongly reduced by introducing only 8% 'big' Y impurities into the LuC layer: RRR = 48 → RRR = 8. The observed approximate plateau for $0.08 \le x \le 0.85$ might be interpreted in terms of strong scattering in the band (group of electrons) with large υ_F, since $H_{c2}(T)$ is only slightly reduced. The relative jump of the electronic specific heat $\Delta C/\gamma T_c \approx 2.06$ at $x = 0.5$ is not very sensitive to impurity effects because it interpolates roughly between the limiting pure limits of 2.27(Lu) and 1.75(Y). The slight reduction of the Sommerfeld constant γ can be ascribed to a broadening of the peak in the DOS.

Some Aspects of Magnetic Borocarbides

The coexistence/competition of magnetism and superconductivity in borocarbides with magnetic rare earth elements is one of the most challenging problems in the field. Most dramatic effects have been observed for $HoNi_2B_2C$, where below the onset of superconductivity at 8.8 K a suppression of superconductivity (SSC) for $\vec{H} \perp$ to the c-axis and $4.5 < T < 5.5$ K has been detected. Three magnetic structures shown in figure have been observed in this region: c-axis modulated commensurate (cc), the spiral c-axis modulated incommensurate (icc) and a-axis modulated incommensurate (ica) ones. From the fact the icc structure and the SSC both occur within a narrow temperature range and that these effects have only been observed for $HoNi_2B_2C$, it has been supposed that the structure is the origin for the SSC. However, replacing Ho partially by the nonmagnetic Y, the magnetic structures for $Y_xHo_{1-x}Ni_2B_2C$ are shifted differently to lower temperatures. This has enabled us to identify the ica phase as the one responsible for the SSC-phenomenon. The observed vector in the a-modulated incommensurate structure Qm = 0.585 is close to the above mentioned calculated nesting vector of 0.6.

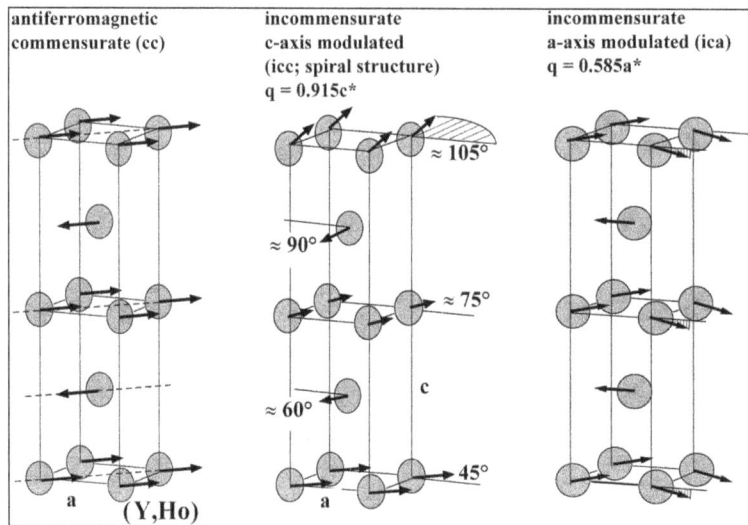

Magnetic structures observed for the $HoNi_2B_2C$ compound.

LuNi₂B₂C and YNi₂B₂C: Unconventional Pairing?

For these two compounds there are several properties, which when taken together, might be interpreted as hints for unconventional (d-wave or p-wave) superconductivity, as follows:

- A $\sqrt{H}$-dependence of the electronic specific heat Cel in the superconducting state instead of the el standard linear dependence.

- The very weak damping of the dHvA oscillations in the superconducting state related to the superconducting gap has been interpreted as strong evidence for a very small or vanishing gap at parts of the Fermi surface.

- A non-exponential non-universal power-like T-dependence of the electronic specific heat in the superconducting phase $C_{el} \alpha T^{\beta}$, $\beta \sim 3$ at low temperatures ($\beta \approx 2.75$ for YNi_2B_2C and $\beta > 3$ for $LuNi_2B_2C$ and $LaPt_2B_2C$). Strictly speaking, a pure one-band d-wave superconductor shows quadratic temperature dependence; a cubic dependence is expected for an order parameter with point-like nodes.

- The anisotropy of $H_{c2}(T)$ within the basal plane of $LuNi_2B_2C$ has been ascribed to d-wave pairing.

- A quadratic flux line lattice at high fields has been observed not only for magnetic RTBC but also for the non-magnetic title compounds.

- Deviations from the Korringa behaviour of the nuclear spin lattice relaxation rate $1/T_1T = $ const have been ascribed to the presence of antiferromagnetic spin-fluctuations on the Ni site.

However, it should be noted that several of these unusual properties have been observed also for some more or less traditional superconductors such as V_3Si and $NbSe_2$. At present it is also unclear to what extent the observed anisotropies could be described alternatively by a full anisotropic multi-band extended s-wave theory. Phase-sensitive experiments and the observation of the Andreev bound state near appropriate surfaces must be awaited to confirm or disprove the d-wave scenario.

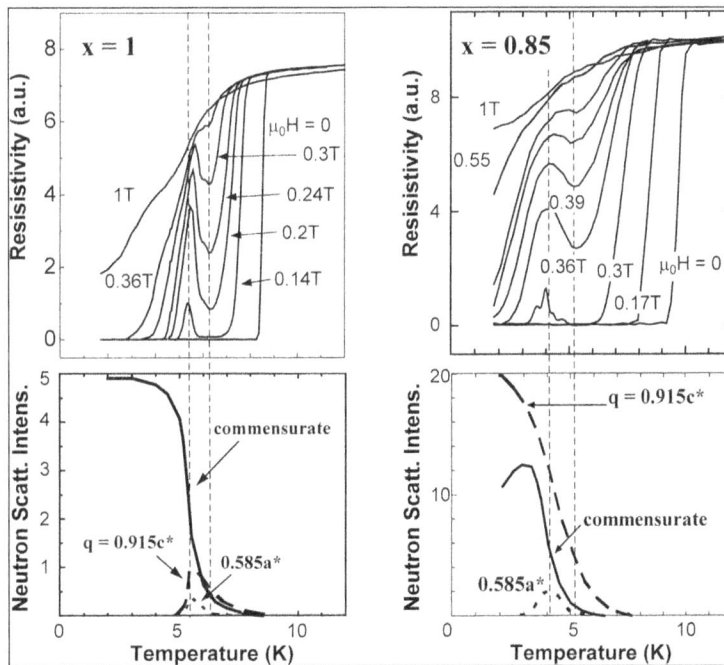

Resistivity vs. temperature for various external magnetic fields for $Ho_xY_{1-x}Ni_2B_2C$ samples with x = 1 and x = 0.85 (top). Intensity of neutron scattering intensity vs. temperature for the magnetic structures.

Most importantly, some of the unusual T-dependences ascribed sometimes to the presence of antiferromagnetic spin fluctuations and 'non-Fermi liquid' effects might be caused by the rather strong energy dependence of the DOS near the Fermi level. Finally, the observation of a weak Hebel–Slichter peak in the 13^C NMR data for the spin-lattice relaxation time $T_1\left(1/T_1T\right)$ must be mentioned. This points to the presence of at least one s-wave pairing component in the multiband order-parameter including also the C2 s-electrons. However, a strong electron–electron interaction is suggested by the twice as large so-called enhancement factor compared with those of conventional 's-band' metals Li and Ag.

In this context, it should be mentioned that according to our LCAO calculations the C2 s contribution to N(0) is very small $(\leq 1\%)$.

Fe-based Superconductors

After the discovery of superconductivity in iron-based superconductors (FeSCs), their very high upper critical fields, low anisotropy and large J_c values, which are only weakly reduced by magnetic fields at low temperatures, suggested considerable potential in large scale applications, particularly at low temperature and high fields. Among the different families, the 122 compounds with a chemical composition of AFe_2As_2 (A = alkaline earth metal) appear to be the most promising, as they are the least anisotropic, have a fairly large Tc of up to 38 K, close to that of MgB_2, and exhibit large critical current densities. However, 122 compounds contain toxic As and reactive alkaline earth metals, which may be a problem for large scale fabrication processes. In this respect, 1111 compounds with the chemical composition LnFeAsO (Ln = Lanthanides) present problems as well, as they contain. As as well as volatile F and O, whose stoichiometry is hardly controlled. 11 compounds with the chemical composition FeCh (Ch = chalcogen ion) have a lower Tc of up to 16 K, but they contain no toxic or volatile elements. It is worth mentioning that new iron-based superconducting families and compounds are regularly discovered, such as for example the 112 compounds (Ca,RE)FeAs$_2$ (RE = rare earth such as La,Ce,Pr,Sm,Eu,Gd) with T_c up to ~40 K, the 42 214 compounds $RE_4Fe_2As_2Te_{1-x}O_4$ with Tc up to ~45 K for RE = Gd, the 21 311 compounds Sr_2MO_3FeAs (M = Sc, V, Cr) with T_c ~ 37 K and [(Li, Fe)OH]FeSe with Tc up to ~40 K.

Thanks to a small coherence length of a few nanometers, FeSCs are particularly sensitive to the inclusion of nanoparticles and to local variation of stoichiometry as pinning centers to enhance the critical current density. For example, the pinning force in 122 films was enhanced above that of optimized Nb_3Sn at 4.2 K by the introduction of self- assembled $BaFeO_2$ nanorods, while similar effects were obtained due to local variations of stoichiometry in 11 films. Critical current J_c values exceeding 105 A cm^{-2} were measured in FeSCs films of 11, 122 and 1111 families up to very large magnetic fields either parallel or perpendicular to the Fe planes. In particular a J_c above 105 A cm−2 was achieved up to 18 T in P-doped $BaFe_2As_2$ films, up to 30 T in $FeSe_{0.5}Te_{0.5}$ films and up to 45 T in SmFeAs(O,F) films. Record values of self-field critical current densities up to 6 MA cm−2 at 4.2 K were measured in 122 films and up to 20 MA cm^{-2} at 4.2 K in zero field in 1111 single crystals irradiated with heavy ions. Furthermore, nanometer scale disorder proved to suppress T_c only very weakly, suggesting that yet further improvements of flux pinning are achievable. In the following, the basic properties of FeSCs and the most important achievements in the development of practical conductors are reviewed with respect to established results on other superconductors. An assessment of the application potential of FeSCs is attempted, based on properties and promising results measured on short specimens. On the other hand, the issue of upscaling preparation procedures must certainly be faced in the future, but iron-based wire technology is currently far less mature than other technologies such as that of $YBa_2Cu_3O_7$ coated conductors.

Table: Relevan iron-based compounds and technical superconductors. The highest T_c found in each respective family is given. Top refers to a typical or expected operation temperature.

Compound	Code	max. T_c (K)	Top (K)	
$LnFeAsO_{1-x}F_x$	1111	58	40	Ln = Sm, Nd, La, Pr,K.
$BaFe_2As_2a$	122	38	25	K, Co, or P doping
$FeSe_{1-x}Te_x$	11	16	4.2	
Nb-Ti	—	10	4.2	
Nb_3Sn	—	18	4.2	
MgB_2	—	39		
$RE-Ba_2Cu_3O_{7-x}$	RE-123	95	77	RE = Y, Gd, Sm, Nd, Yb,K
$Bi_2Sr_2CaCu_2O_{8-x}$	Bi-2212	85	20	
$Bi_2Sr_2Ca_2Cu_3O_{10-x}$	Bi-2223	110	77	

Basic Properties

The basic properties set the final performance limit of a superconducting material in terms of temperature, field and critical current.

Transition Temperature

The highest transition temperature of all FeSCs was found in the 1111 compound $T_c \sim 58$ K, which places this compound between the cuprates and MgB_2. The transition temperature T_c is defined as the temperature up to which superconductivity persists. However, applications are restricted to lower temperatures, since superconductivity becomes very weak close to T_c. As a rule of thumb, the operation temperature Top should be about half of T_c or lower in applications requiring high currents and/or fields. However, strong thermal fluctuations of the vortex lattice reduce the critical currents significantly in highly anisotropic materials, restricting appropriate operation conditions to much lower temperatures. $(Bi,Pb)_2Sr_2Ca_1Cu_2O_x$ (Bi-2212) is an extreme example that provides useful current densities only at temperatures below about 20 K, despite its high transition temperature of 85 K. The anticipated maximum operation temperatures of the FeSCs are given in table together with values for other superconducting compounds. Note that the higher the required magnetic field, the lower the operation temperature must be. Since the 1111 compounds are the most anisotropic of all considered FeSCs, the estimated value for the maximum Top has to be confirmed when long length conductors become available, and may be restricted to low magnetic fields. In any case, all FeSCs dis- covered so far are obviously no alternative to $ReBa_2Cu_3O_7$(RE-123) coated conductors or $(Bi,Pb)_2Sr_2Ca_2Cu_3O_x$ (Bi- 2223) tapes at high temperature (>50 K), in particular for use with nitrogen as the coolant.

K-doped BaFe$_2$As$_2$ (Ba-122) has a transition temperature of around 38 K, nearly the same as MgB$_2$. From this view- point, the two materials are direct competitors for applications at intermediate temperatures, which do not rely on liquid helium as a coolant. P- or Co-doped Ba-122 have lower Tcs of about 30 K and 24 K, respectively, which makes helium free operation questionable. The 11 compounds have the lowest T$_{cs}$, even below that of the readily available Nb$_3$Sn, thus helium cooling is the only option.

Although many new iron-based compounds have been discovered, T$_c$ has not significantly increased over the past few years. However, superconducting-like energy gaps between 55 K and 75 K were found by angle resolved photoemission spectroscopy (ARPES) inspections in a single layer of FeSe on top of doped SrTiO$_3$, possibly resulting from a charge transfer between the superconducting layer and the substrate. Unambiguous evidence of a transition to zero resistivity and diamagnetic behavior is yet to be widely established in this system at such high temperatures, although a highly non-linear behavior in the I–V curves up to about 100 K was derived from a particular in situ four-point probe technique, which exactly fits the expected behavior of a superconductor. The temperature and field dependencies of the derived Jc and resistivity are also consistent with superconductivity. These results fuel the hope that higher Tcs are achievable in iron- based compounds with as yet unknown interlayers.

Upper Critical Field

The upper critical field B$_{c2}$ limits the field which can be generated using the respective superconductor; maximally about 0.75. B$_{c2}$ can be effectively achieved. Since superconducting wires are used nowadays nearly exclusively for magnets, B$_{c2}$ is certainly a key parameter for applications and restricts available magnets based on conventional (niobium- based) technology to fields below 25 T. Novel conductors for the next generation of NMR, accelerator, research, and fusion magnets are urgently needed. While MgB$_2$ is unsuitable for high field magnets, cuprates and FeSCs have upper critical fields in the 50–100 T range (and even greater) at 4.2 K, thus not imposing any realistic limitations for high field magnets operating at low temperatures. On the other hand, B$_{c2}$ decreases with temperature and converges to zero at T$_c$, thus a high B$_{c2}$(0K) is, besides a high-T$_c$, a prerequisite for cryocooled magnets operating at intermediate fields (e.g. medical MRI magnets). In this respect, FeSCs are clearly favorable compared to MgB$_2$, which is already applied in low field MRI systems.

An important point to mention is the low anisotropy of the upper critical field $B_{c2}^{(ab)} / B_{c2}^{(c)}$ in the FeSCs, which makes flux pinning more efficient than in the highly anisotropic cuprates by reducing flux cutting effects and thermal fluctuations. In particular, the 11 and 122 compounds are nearly isotropic at low temperatures. Although the anisotropy increases with temperature in these compounds, reaching values up to about 3 close to T$_c$, it remains well below that of RE-123 coated conductors ($\approx$5) and Bi-tapes (>20), also at high temperatures.

Critical Current Densities

The critical current density in a superconducting wire is either limited by flux pinning or granularity. Flux pinning is an extrinsic property, which can be tuned by generating a suitable defect structure. The maximally achievable loss free currents are, however, not independent from the basic material parameters, since J_c amounts to maximally 10–20% of the depairing current density, J_d, in optimized materials. J_d is a material property and can be estimated from the coherence length ξ and London penetration depth λ as $J_d \sim \phi_0 \sqrt[3]{3}\pi\mu_0\lambda^2\xi$ (0 is the flux quantum and μ_0 the vacuum permeability). J_d values in the zero temperature limit can reach up to $3 \cdot 10^8$ A cm^{-2} in cuprates, about $1.8 \cdot 10^8$ A cm^{-2} in Nb$_3$Sn (assuming ξ = 3.6 nm and λ = 124 nm) and $2 \cdot 10^8$ A cm^{-2} in MgB$_2$. It turns out similar in SmFeAsO$_{1-x}$F$_x$ and K-doped Ba-122 (about $1.7 \cdot 10^8$ A cm^{-2}), but smaller in the P- and Co-doped 122 system (≈ 5 and $9 \cdot 10^7$ A cm^{-2}, respectively) and only around $2 \cdot 10^7$ A cm^{-2} in the 11 system. Thus at least some compounds can compete with the high values in cuprates, MgB$_2$, and Nb$_3$Sn.

Efficient pinning can be realized comparatively easily in the iron-based materials, as demonstrated by irradiation experiments, by the successful introduction of nanoparticles or nanorods, by the effect of local variation of stoichiometry. Moreover, irradiation with Au ions and neutrons and introduction of artificial ab plane pins emphasized that the introduction of pinning defects does not affect T_c appreciably. This indicates that FeSCs tolerate a higher density of defects without a significant decrease in Tc than cuprates, which makes them ideal candidates for high field applications, since the number of pinning centres is of crucial importance at high fields.

Another key property of FeSCs relevant for applications is the small anisotropy of J_c with respect to the crystal axis. Direct transport J_c measurements in the two main crystallographic directions $J_c^{(ab)}$ and $J_c^{(ab)}$ were carried out on Sm-1111 single crystals with patterned micro-bridges and on Ba(Fe$_{1-x}$Co$_x$)$_2$As$_2$ single crystals using the Montgomery technique. The obtained $J_{c2}^{(ab)}/J^{(c)}$ $J_c^{(ab)}/J_c$ (c) ratios were 2.5 and 1.5 respectively, much lower than the values of up to 10–50 found in the cuprates.

Grain Coupling

All high-T_c superconductors are prone to magnetic granularity, which limits the macroscopic currents. While secondary phases residing at the grain boundaries and voids reduce the cross section over which the current effectively flows in MgB$_2$, high angle grain boundaries intrinsically limit the currents in untextured polycrystalline cuprates. For misalignment angles between adjacent grains above $\Theta c \sim 3°$, Jc drops exponentially. Unfortunately, such an exponential decay of the current as a function of the misalignment angle between grains was measured in the FeSCs as well, namely in 122 films grown onto bicrystal substrates. However, the suppression of Jc is not as strong as in high-T_c cuprates; indeed it was found that the critical angle for J_c suppression is slightly larger than in cuprates $\Theta c \sim 9°$ and the suppression itself is less severe, for example for Θ from 0° to 24° Jc decreases by one order of magnitude in Ba(Fe$_{1-x}$Co$_x$)$_2$As$_2$ and by two orders of magnitude in YBa$_2$Cu$_3$O$_{7-x}$. On the other hand, it was suggested that 'real' grain boundaries can often show much better transparency than the planar grain boundaries of the bicrystals, because the misorientation angle is not the only parameter that determines whether or not grain boundaries are transparent to the supercurrent. In addition, the orientation of the field with respect to the grain boundary has to be taken into account, because the inter-grain J_c degrades the most when a significant portion of the vortex lies in the grain boundary. When the vortex obliquely crosses the grain boundary, the suppression of the inter-grain J_c is much weaker.

On the whole, it can be stated that the weak link problem is less serious in FeSCs than in cuprates. The mechanisms that limit current flow at the grain boundaries in FeSCs are still lacking a well-founded explanation. There are likely a number of reasons, both intrinsic and extrinsic, such as the larger critical angle Θ_c, possibly related to the higher robustness of the superconducting s-wave symmetry as compared to d-wave symmetry in cuprates, and the metallic nature of underdoped phases that may be present at the grain boundaries as compared to the insulating nature of cuprate parent compounds.

Magnetic granularity in the cuprates was (at least partly) overcome by texturing in the coated conductor technology. A high degree of texture ensures a small density of high angle grain boundaries that would reduce the macroscopic current. The corresponding production techniques, however, involve multiple steps with related costs.

Texturing might not be necessary for the FeSCs. In particular, results on K-doped 122 wires are encouraging, since current densities that approach the requirements of applications have been demonstrated. A combination of the more favorable grain coupling and nanosized grains enable current densities in granular iron-based materials, which are orders of magnitude higher than the best results achieved in untextured cuprates.

Overdoping has a beneficial effect on the inter-grain transport in cuprates, specifically Ca doping in Y-123. Analogously, it was found that Sn addition largely improves inter-grain connectivity and thus Jc in $SmFeAs(O_{1-x}F_x)$, which exceeds $1 \cdot 10^4$ A cm^{-2} at 5 K. In addition, intergrain J_c of $Ba(Fe_{1-x}Co_x)_2As_2$ sintered bulk is enhanced by applying low temperature reaction and Co-overdoping up to x = 0.12, resulting in a similarly high inter-grain J_c at 5K.

Conductor Development

The fabrication of conductors for power applications has been explored since the very beginning of the research activity on FeSCs. The current state-of-the-art is not yet mature enough to address the systematic fabrication of long length specimens, but very encouraging results have been obtained on short samples fabricated both by the powder-in-tube (PIT) method and by processes which replicate the RE-123 coated conductor technology.

Powder-in-tube Processed Conductors

Wires and tapes of all three main FeSC families have been fabricated so far. The quite isotropic character of J_c with respect to the crystalline direction and better coupling between misaligned grains suggests that texturing may not be as stringent as in cuprates, and conductors in the form of untextured wires may achieve the required performance. The fabrication method is PIT, which starts by packing powders into a metallic tube in a high purity Ar atmosphere and sealing the ends. Either powders of the already reacted superconducting phase (ex situ PIT) or powders of precursor phases (in situ PIT) are used, however for 122 and 1111 families, the ex situ process has been used almost exclusively, as it offers more options for optimizing the powder reaction, even by multiple steps. Indeed, powder preparation is crucial for the final result and involves a proper choice of stoichiometry to compensate for element losses during the whole process, the use of high purity precursors, and ball-milling to obtain a smaller grain size and enhance the packing density. The metal tube forming the sheath is generally made of Ag in the case of 122 and 1111 wires and tapes, while a different situation occurs in the case of in situ PIT wires and tapes of the 11 phase, where a Fe sheath is employed, as discussed in the following. Indeed, Ag does not react significantly with the superconducting phase at the optimized temperatures of the final thermal treatments and is thus preferred over Ta, Nb, Cu and Fe. Ag may be also used in combination with an additional outer sheath made of Fe, Ni or stainless steel to reduce costs and improve the mechanical strength. In this respect, very recently, copper sheathed 122 tapes were fabricated with transport $J_c \sim 3.1 \bullet 10^4$ A cm^{-2} at 10 T, which is noteworthy, given that the use of copper as a sheath material is cost effective, has good mechanical properties and provides reliable thermal stabilization in a magnet during transients. This result was obtained with a hot pressing process where the annealing time was minimized, thus inhibiting the formation of a reaction layer at the copper/superconductor interface. The subsequent step of the PIT process is the deformation, carried out by drawing, groove rolling or flat rolling. Finally, thermal treatments are carried out to form the final phase (in situ process) or heal cracks induced by the deformation and enhance grain connectivity and density. The latter goal may be also pursued by performing thermal treatments under high uniaxial pressures. Numberless variations for the deformation and thermal treatment steps are possible and a steady optimization is currently underway.

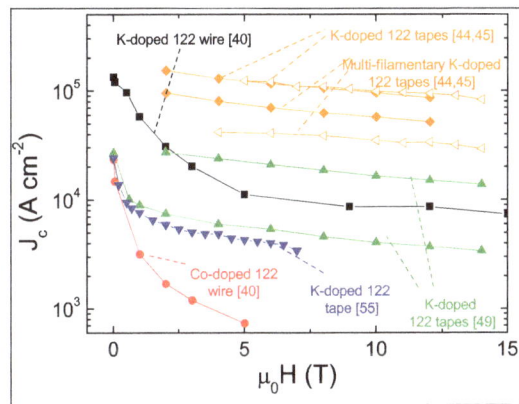

Transport critical current densities at 4.2 K as a function of applied magnetic field. Data refer to K-doped 122 textured tapes with and without chemical additions, prepared by flat rolling and different heat treatments, uniaxially mono- and multifilamentary K-doped 122 tapes prepared using GPa uniaxial pressure, mono- and multi-filamentary K-doped 122 tapes prepared using hot pressing at ~30 MPa and 850-900 °C (open symbols), a K-doped 122 untextured wire containing a high density of grain boundaries and a Co-doped untextured wire with a lower density of grain boundaries.

The best transport critical current values among ironbased superconductor wires and tapes, exceeding 10^5 A cm^{-2}, have been obtained with the 122 family so far. A set of $J_c(H)$ curves measured in 122 wires and tapes, showing the best and most representative behaviors, is presented in figure.

Several routes are studied to improve Jc(H) performances. Chemical addition is well-established for improving grain crystallization and enhancing the metallic character of secondary phases at grain boundaries, thus promoting intergrain coupling. It was shown that 10%–20% Ag substitution significantly improves inter-grain connectivity and reduces porosity of 122 tapes, resulting in a three times increase of the global J_c up to high fields. Pb addition, on the other hand, promotes grain growth and improves grain connectivity for a content up to 10%, yielding an enhancement of the global J_c in the low field regime. Sn addition has a similar beneficial effect in 122.

Nonetheless, with respect to the role of chemical additions for promoting grain growth, it must be remarked that grain growth is not necessarily the right method to pursue. Indeed, it was shown that untextured polycrystalline $(Ba_{0.6}K_{0.4})Fe_2As_2$ bulks and round wires with high grain boundary density, i.e. small grains obtained by ex situ PIT and low temperature (600 °C) thermal treatment, have transport critical current densities well above 10^5 A cm^{-2} in a selffield at 4.2 K. The enhanced grain connectivity was ascribed to significantly improved phase purity. The effectiveness of high density of GBs as pinning centers was explained in terms of low anisotropy and consequent enhanced vortex stiffness of 122 compounds.

Densification of the superconducting core by uniaxial pressing under a high pressure of ~2 GPa before sintering yielded transport J_c values above 10^5 A cm^{-2}, still as high as $8.6 \cdot 10^4$ A cm^{-2} at 10. This result can be attributed to a change in the crack structure and a more uniform deformation achieved by pressing rather than rolling. Indeed, as already observed in Bi-2223 tapes, cracks run transversely to the tape length for rolled tapes and parallel to the tape length for pressed tapes. An optimization of the cold deformation process has to balance the improvement in density and the initiation of microcracks, which cannot be healed by a subsequent heat treatment.

A further improvement in this direction was achieved by a hot pressing technique, which makes the grains more flexible to couple with each other without producing a large number of crashed grains, thus significantly reducing the voids and cracks and leading to a denser superconducting core of PIT tape. More specifically, by pressing K-doped 122 tapes at ~30 MPa and high temperature THP, better homogeneity, texture and grain connectivity were obtained, yielding a transport J_c above 10^5 A cm^{-2} at 10 T for THP = 850 °C and even above 10^5 A cm^{-2} at 14 T for THP = 900 °C. However, it should be pointed out that the practical application of uniaxial pressing for the manufacture of long length wires requires specialized machines for continuous pressing of the tape. An easy and simple process is required to balance the high performance and the production cost of the superconducting tapes. So far, through scalable and cheap processes, ~12 cm long iron-based superconducting tapes with high transport J_c ($\sim5.4 \cdot 10^4$ A cm^{-2} $- 7.7 \cdot 10^4$ A cm^{-2} at 10 T and 4.2 K) have been obtained. In addition, the fabrication by rolling of a remarkable 11 m long 122 tape, exhibiting fairly uniform J_c throughout the sample, always above $1.5 \cdot 10^4$ A cm^{-2} at 10 T and 4.2 K, has been reported. In scalable processes, mechanical deformation by groove and flat rolling is critical, as its beneficial role in densifying the conductor core and aligning the grains is counterbalanced by the detrimental effect of current blocking transverse cracks generated in the microstructure. Annealing treatments only partially heal microcracks, so that an optimized procedure may require multiple incremental steps of repeated rolling and heat treatment.

Improved texture in general improves transport J_c, with values above 10^4 A cm^{-2} up to fields as high as 14 T. Indeed, at high fields, textured tapes perform better than the best untextured wires. The beneficial effect of texturing fuelled the idea of preparing iron-based superconductor coated conductors. Texturing certainly helps to overcome the problems related to anisotropy, but these problems should not be as severe as for high-Tc cuprates, given the smaller values of $J_{c2}^{(ab)} / J_c^{(c)}$ and $B_{c2}^{(ab)} / B_{c2}^{(c)}$ anisotropies found in FeSCs. In addition, the problems related to the current blocking effect resulting from the depressed superconducting order parameter at large angle grain boundaries may have a less important role in FeSCs than in the cuprates. Indeed, as mentioned above, granular iron-based materials exhibit better grain coupling than untextured cuprates, demonstrated by their current densities, which are higher by orders of magnitude than the best results in untextured cuprates. This may be explained also by the clean nature of grain boundaries obtained by an optimized thermal treatment, which drastically reduces FexAsy secondary phases in the 122 compounds.

In fact, a great effort has been carried out with the goal of finding the optimized temperature of thermal treatment T_{tt} to enhance the transport critical current. In general, for tapes, T_{tt} values in the range 700–900 °C are used, while in the case of wires a temperature T_{tt} = 600 °C has allowed the best transport J_c(H = 0) values so far. By comparing the preparation conditions of the tapes exhibiting the best self-field and high field transport critical current density, it turns out that with increasing thermal treatment temperature c-axis texture, crystallinity and grain connectivity are improved, but eventually secondary FexAsy phases appear and, in the case of hot uniaxial pressing, transverse microcracks also develop. As a consequence, an optimal T_{tt} can be identified. From the data collection shown in figure, it can be gathered that, in terms of maximum self-field transport J_c, such an optimized T_{tt} is 850 °C, regardless of the fabrication either by simple flat rolling or by further application of ~GPa uniaxial pressure. On the other hand, by assessing the high field J_c(H ~ 10 T) as a quality parameter, T_{tt} = 850 °C is again the best T_{tt} value if the fabrication is carried out with no external applied pressure, while if the process involves application of ~GPa uniaxial pressure the best J_c(H ~ 10 T) results are obtained using T_{tt} = 900 °C. This difference likely arises from the thermodynamic conditions for secondary phase formation and from the lower volatility of elements under high pressure, which preserves the correct stoichiometry up to higher temperatures. It must be remarked that in some cases, even if no external pressure is applied, the thermal treatment is carried out in sealed stainless steel tubes, which also inhibits the loss of volatile elements. On the other hand, if the thermal treatment is carried out in flowing Ar, following a truly scalable process, the optimal Ttt is found to be slightly smaller, namely T_{tt} = 800 °C. Note that T_c and J_c are not exactly optimized in the same range of T_{tt} values. The T_c of $Sr_{0.6}K_{0.4}Fe_2As_2$ tapes increases with increasing heat treatment temperature, while maintaining a narrow transition width, up to 900 °C. This behavior is explained by improved crystallinity while approaching the synthesis temperature of precursors, before secondary phases eventually form at T_{tt} > 900 °C.

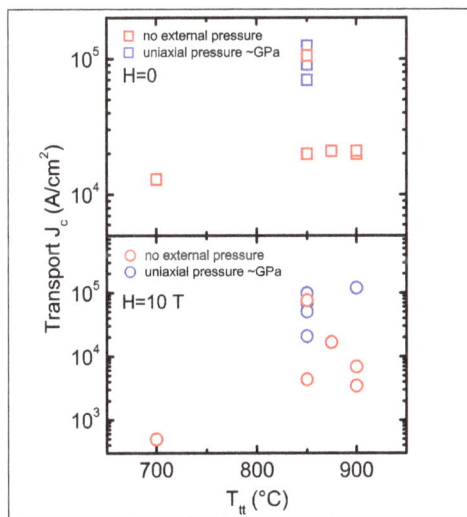

Transport critical current densities of 122 tapes at 4.2 K in zero (upper panel) and 10 T (lower panel) magnetic field plotted as a function of the thermal treatment temperature T_{tt}. Data in zero and high field of samples prepared with no applied external pressure are taken from. Data in zero field of samples prepared at high pressure are taken from. Data at 10 T of samples prepared at high pressure are taken from.

Another key parameter to optimize is the applied pressure, either applied via deformation processes or via isostatic ~GPa uniaxial pressing techniques. Indeed, increasing pressure favors densification and grain connectivity, however it eventually yields to the formation of current blocking microcracks. This parameter does not show a clear tendency in the literature.

As for multi-filamentary 122 iron-pnictide wires and tapes, the highest transport Jc values reached so far are $6.1 \cdot 10^4$ A cm^{-2} and $3.5 \cdot 10^4$ A cm^{-2} at 4.2 K and 10 T, respectively for hot pressed 7- and 19-core Sr-122 tapes. A very recent work on 7-, 19- and 114-filament $Sr_{1-x}K_xFe_2As_2$ wires and tapes indicated that J_c is enhanced by improving densification and filament microstructural uniformity along the longitudinal direction. Both these targets can be pursued by optimizing the cold deformation process in thinner tapes with a lower number of filaments. Less work has been carried out for the fabrication of 1111 wires and tapes, due to the difficulty in controlling O and F stoichiometry during heat treatments at high temperatures. The commonly used sintering temperature for 1111 wires is 1200 °C, but sintering at 850 °C–900 °C yielded similarly high-T_c and transport J_c up to 1300 A cm^{-2}.

Low temperature sintering allows one to prevent FeAs liquid phases from forming. On the other hand, many unreacted precursors remain in the 1111 bulks even after long sintering, as the reaction rate at low temperatures is relatively small.

Moreover, tapes cannot endure long sintering without fluorine loss. In addition, in the case of 1111 wires and tapes, Ag was found to be the best sheath material and metal additions were found to be beneficial. Indeed, the loss of fluorine is reduced and the intergranular coupling enhanced in Sn-added tapes prepared by ex situ PIT, exhibiting transport J_c as large as $2.2 \cdot 104$ A cm^{-2} at 4.2 K in a self-field. Pre-sintering of Sn-added powders allows one to reduce the FeAs wetting phase and fill the voids between Sm-1111 grains, yielding improved grain connectivity and transport J_c up to $3.45 \cdot 10^4$ A cm^{-2} at 4.2 K in a self-field. However, J_c rapidly decreases with increasing magnetic field, dropping to around 102 A cm^{-2} at 8 T possibly as a consequence of the larger anisotropy of the 1111 compounds as compared to the 122 family.

Wires and tapes of the 11 family are appealing for applications as well. Indeed the 11 family, despite its lower T_c, exhibits high J_c and H_{c2} and does not contain toxic elements. Fe(Te,Se) polycrystals prepared by combined melting and annealing processes exhibit a highly homogeneous and dense microstructure, characterized by large and well interconnected grains. A global critical current density, reaching about 10^3 A cm^{-2}, was measured in these samples.

Despite this encouraging starting point, several difficulties were encountered in the fabrication of 11 wires and tapes, due to chemical reactions between the superconductor and the sheath during thermal treatments and difficulties in obtaining a high density of powder inside the tube. For this reason, Fe turned out to be the best choice for the sheath, as it allows a diffusion process, where Fe-free precursors are sealed inside Fe tubes and the final 11 phase is formed by the supply of Fe from the sheath during the thermal treatment. This diffusion process yielded a significant improvement in the transport J_c reaching values up to 10^3 A cm^{-2} in FeSe wires. As compared to FeSe, the $FeTe_{0.5}Se_{0.5}$ compound exhibits better superconducting properties in terms of T_c, J_c and H_{c2}.

However, only transport J_c values around 220 A cm^{-2} and 400 A cm^{-2} were measured in FeTe$_{0.5}$Se$_{0.5}$ wires. Indeed, the preparation of Fe(Te,Se) wires and tapes is more difficult. It was found that the starting powders de-compose to a Fe$_{1+y}$(Se,Te) phase with Se/Te $\approx$ 1 and excess Fe plus a FeSe$_{1-y}$ phase after the heat treatment. The former phase is not superconducting, due to the excess Fe which is detrimental to superconductivity, while the latter phase is superconducting with $T_c \sim 8$K, much smaller than the $T_c \sim 16$ K of FeTe$_{0.5}$Se$_{0.5}$. Regarding the optimization of the thermal treatment, it was found that the superconducting properties of FeSe wires improve with increasing annealing temperature up to 1000 °C, where the phase formation is complete. A proper thermal treatment of ex situ PIT Fe(Te,Se) wires allows one to enhance the packing density of the core inside the sheath, thanks to the expansion of the lattice volume during the transformation from high density hexagonal Fe(Te$_{0.4}$Se$_{0.6}$)1.4 to low-density tetragonal FeTe$_{0.4}$Se$_{0.6}$, thus yielding enhanced magnetic J_c up to $3 \cdot 10^3$ A cm^{-2} at 4.2 K in self-field.

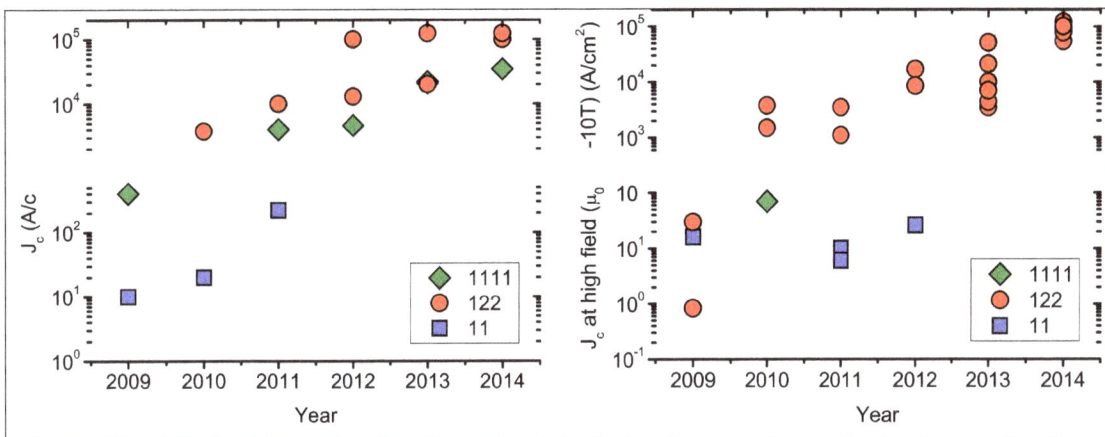

Transport Jc values obtained in iron-based polycrystals, wires and tapes versus the year of publication. In the left panel, data at T = 4–5 K and self-field are reported while data at T = 4–5 K and a field of 8–10 T are shown in the right panel.

In summary, the highest transport critical current in ironbased superconducting wires and tapes have so far been obtained with the 122 family, namely up to 104–105 A cm^{-2}. Moreover, in 122 wires the Jc field dependence is quite flat, with a decrease of one order of magnitude from a self-field to a field well above 10 T. For the 1111 family, the transport Jc values found in wires and tapes prepared by ex situ PIT reach $3.45 \cdot 10^4$ A cm^{-2}, but the field dependence of J_c is steeper as compared to 122 wires and tapes. Wires and tapes of the 11 compounds obtained by in situ PIT exhibit the lowest transport Jc values, up to $3 \cdot 10^3$ A cm^{-2}, but they have the advantages of containing no toxic arsenic and having the simplest crystal structure. The results achieved so far seem to indicate that among the key targets to pursue are (i) improving densification and (ii) inducing a certain degree of texture. Simultaneously, key issues to avoid, not unique to FeSCs, are the segregation of large precipitates at the grain boundaries and the formation of cracks during the deformation process. The best J_c performances are obtained with thermal treatments under high uniaxial pressures, which result in both enhanced density and texture. Hence a finely tuned multi-step protocol of mechanical deformation plus thermal treatment could be set up for tape fabrication, whose effects mimic those of thermal treatments under high uniaxial pressures, minimizing the presence of cracks and fulfilling the further requirement of scalability. However, this would be at the expense of a less favorable geometry, namely tapes instead of wires, the latter being much more favored by engineers. As no evidence of local texture in round conductors, as in the case of melt textured Bi-2212 wires, has ever been detected in FeSCs so far, a possible route could be restacking the tapes into a tube and drawing a wire, similar to the Bi-2212 ROSAT wires. Alternatively, the performances of untextured wires could be improved by grain size refinement, which effectively enhances intergrain coupling.

From the state-of-the art results, it can be envisaged that iron-based superconductor (122) wires and tapes are promising for magnet applications at 20–30 K, where the niobium- based superconductors cannot play a role owing to their lower T_c, and J_c being rapidly suppressed by the applied field in MgB_2. Moreover, the steady improvements of J_c values achieved in wires and tapes based on 11, 122 and 1111 ironbased superconductors during recent years hardly seem to be saturating yet. This positive trend suggests, on one hand, that the inter-grain current in real conductors may behave better than that in epitaxially grown bicrystals. On the other hand, there exists a considerable potential for future improvements of these materials.

Coated Conductors

The evidence of weak-linked behavior in 122 thin films grown onto bicrystals suggested the application of the coated conductor technology to iron-based superconductors, i.e. depositing iron-based superconductor films on textured metal substrates with buffer layers, by the techniques successfully developed for second-generation cuprate wires. Ion-beam assisted deposition (IBAD) coated conductor templates are manufactured in two steps. First, an Y_2O_3 layer is made on unpolished Hastelloy by sequential solution deposition to reduce the roughness of the tape surface.

Then a biaxially textured MgO layer is deposited on top by IBAD. Through this technique, biaxial texture is achieved by means of a secondary ion gun that orients the oxide film buffer layer while it is being deposited onto the polycrystalline metallic substrate. Alternatively, the RABiTS process for coated conductor templates achieves texture by mechanical rolling of a face-centered cubic Ni–W alloy and subsequent heat treatment. A series of biaxially textured buffer oxides, such as Y_2O_3, YSZ and CeO_2 is grown on such metal substrates.

122 films have been grown on IBAD substrates. In-plane misorientation of 3°–5° was measured and, most importantly, J_c values of 10^5–10^6 A cm^{-2} were achieved. This route turned out to be encouraging for the 11 family as well. Fe(Se,Te) thin films deposited on IBAD-MgO-buffered Hastelloy substrates were able to carry transport critical current up to 2 • 105 A cm^{-2} at low temperature and self-field, still as high as 104 A cm^{-2} at a field of 25 T. Even more remarkable results were obtained for Fe(Se,Te) thin films deposited on RABiTS substrates, namely critical currents up to 2•10^6 A cm^{-2} at low temperature and self-field, still as high as 10^5 A cm^{-2} at a field of 30 T. The fabrication of coated conductors with 1111 FeSCs was also attempted. NdFeAs(O,F) thin films grown by molecular beam epitaxy on IBAD-MgO-Y_2O_3 Hastelloy substrates showed a high c-axis texture, but not complete in-plane texture. A magnetic J_c of 7 • 10^4 A cm^{-2} was measured in a self-field at 5 K, which is larger by one order of magnitude than the J_c of 1111 PIT tapes, but significantly smaller than the J_c of 122 and 11 coated conductors. In addition, the field and temperature dependence of J_c is much stronger than that of coated conductors of the other families, as a consequence of the weak link behavior related to incomplete biaxial texture and the higher anisotropy of this compound.

Remarkably, the challenging fabrication of long (>1m) coated conductors using a pulsed laser deposition system equipped with a reel-to-reel tape feeding mechanism has already been undertaken.

P substituted Ba-122 films were deposited on IBAD-MgO-buffered Hastelloy tapes. J_c measurements on 5 cm and 10 cm pieces yielded $1.1 \cdot 10^5$ A cm^{-2} and $4.7 \cdot 10^4$ A cm^{-2}, respectively. Problems of P loss in the film and inhomogeneity over long lengths were identified.

In the assessment of the potential of coated conductors, it must be pointed out that the thickness of the superconducting layers in coated conductors is currently less than 150 nm. Thicker layers have not been attempted so far but are not expected to be difficult to achieve. Considering the very low engineering critical current density Je (ratio of critical current I_c to the whole cross-sectional area of conductor), no advantageous points can be found for the iron-based coated conductors compared to the PIT processed tapes and wires at the present stage. Moreover, higher production cost and low production rates of coated conductor technology must be taken into account when assessing the application potential of this process as compared to that of the PIT technology.

However, the elaborated oxide buffer structure, partially designed to protect the metal template from oxidation for cuprates wires may not be needed at all for Fe(Se,Te) wires deposited in a vacuum. Growing a Fe(Se,Te) coating directly on textured metal tapes may be possible, thus greatly simplifying the synthesis procedure, reducing production costs and avoiding possible uncontrolled oxygen diffusion into the intermetallic superconductor.

References

- Gap-structures-of-a-15-alloys-from-the-superconducting-and-normal-state-break-junction-tunnelling, superconductors-new-developments: intechopen.com, Retrieved 24 March, 2019

- Heavy-fermion-superconductors, physics-and-astronomy: sciencedirect.com, Retrieved 23 July, 2019

- Heavy Fermions: Electrons at the Edge of Magnetism. Handbook of Magnetism and Advanced Magnetic Materials (2007)

- 8 Ironbased layered superconductor LaO1−xFiFeAs (x = 0.05−0.12) with Tc = 26 K J. Am. Chem. Soc. 130 3296

Applications of Superconductors

Superconductors have a wide range of applications in the fields of power industries, automobiles, medicine, etc. They are also used in maglev trains, magnetic resonance imaging and nuclear magnetic resonance machines. These diverse applications of superconductors have been thoroughly discussed in this chapter.

Application in Power Industry

Application of superconductors in power mainly refers to their attractive feature in transmitting the energy without loss. Based on this fact various field of application are being discovered for super-conductors and newer fields are added continuously.

Superconducting Wire

Superconducting wire or cable is made of superconducting materials which, when cooled below its transition temperature, has zero electrical resistance. Often the superconductor is in filament form or on a flat metal substrate encapsulated in a copper or aluminum matrix that carries the current should the superconductor quench (rise above critical temperature) for any reason.

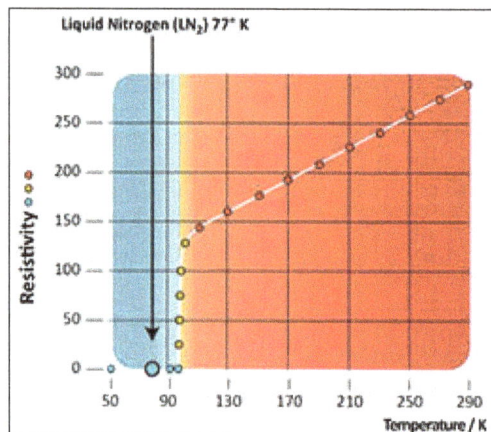

Superconducting Cable Design

The so called 'warm dielectric design' is based on a flexible support with stranded HTS tapes in one or several layers forming the cable conductor. This conductor, cooled by the flow of liquid nitrogen, is surrounded by a cryogenic envelope employing two concentric flexible stainless steel corrugated tubes with vacuum and superinsulation in between. The structure of a superconducting cable is shown in figure.

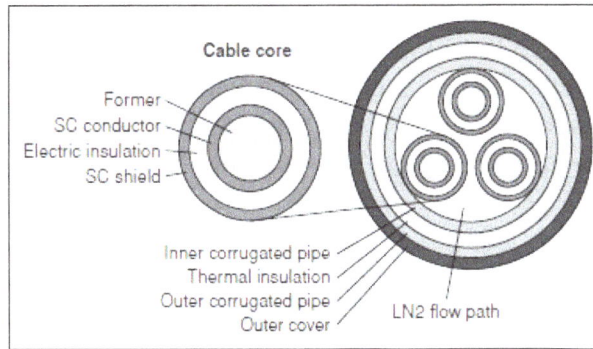

Structure of superconductor cable.

The conductor is formed by laying the Bi-based superconducting wires in a spiral on a former. Polypropylene Laminated Paper (PPLP) is used for the electrical insulation due to its good insulation strength and low dielectric loss at low temperatures, and liquid nitrogen works as a compound insulation in addition to coolant. On the outside of the insulation layer, a superconducting wire of the same conductor material is wound in a spiral to form a shield layer. Each shield layer of each core is connected to each other at both end of the cable, so that an electrical current of the same magnitude as that in the conductor is induced in the shield layer in the reverse direction, thus reducing the electromagnetic field leakage outside the cable to zero. Three cores are stranded together, and this is placed inside a double-layered SUS corrugated piping. Thermal insulation is placed between the inner and outer SUS corrugated piping's, where a vacuum state is maintained to improve the thermal insulation performance. A schematic view is shown in figure.

Warm dielectric superconducting cable.

With the outer dielectric insulation, the cable screen and the outer cable sheath are at room temperature. Compared to conventional cables, this design offers a high power density using the least amount of HTS-wire for a given level of power transfer. The second type of HTS cable design is the 'cold dielectric' as shown in figure, using the same phase conductor as the warm dielectric, the high voltage insulation now is formed by a tape layered arrangement impregnated with liquid nitrogen (LN_2).

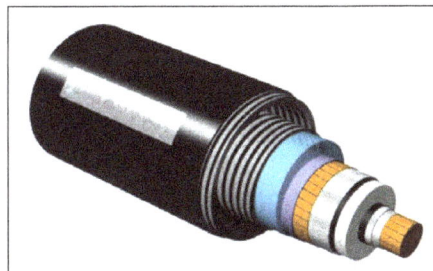

Cold dielectric superconducting cable.

Thus LN_2 is used also as a part of the dielectric system in the cold dielectric cable design. The insulation is surrounded by a screen layer formed with superconducting wires in order to fully shield the cable and to prevent stray electromagnetic field generation. Three of these cable phases can either be put into individual or, into a single cryogenic envelope. A special type of cold dielectric cables is represented by the 'triaxial design', shown in figure.

Triaxial cold dielectric cable.

This design has three phase conductors concentrically arranged on a single support element divided by wrapped dielectric, which has to withstand the phase-to-phase voltage. The differences in superconducting power cable designs have significant implications in terms of efficiency, stray electromagnetic field generation and reactive power characteristics. In the warm dielectric cable design no superconducting shield is present, thus no magnetic shielding effect can be expected during operation. As a consequence, higher electrical losses and higher cable inductance are significant drawbacks relative to the other superconducting cable designs. Additional spacing of the phases is necessary in warm dielectric configuration due to electrical losses influenced by the surrounding magnetic field. No such requirements exist for cold dielectric cables, as shown in Table, a calculated example compared to conventional cables and overhead lines. On the other hand, the warm dielectric configuration seems easier to achieve as many components and manufacturing processes are well-known from conventional cables. The cold dielectric cable is more ambitious as it involves new developments in the field of dielectric materials and also needs more complicated cable accessories, such as terminations or joints.

Table: Electrical Characteristics Example of 120 kV Class Cable.

A Comparison of Power Transmission Technologies			
Technology	Resistance (Ω/km)	Inductance (mH/ km)	Capacitance (nF/km) (MVAR/km)
Cold Dielectric HTS	0.0001	0.06	200/1.08
Conventional XLPE	0.03	0.36	257/1.40
Overhead Line	0.08	1.26	8.8/0.05

High Temperature Superconducting (HTS) Wire

There are two well-recognized types of high temperature superconducting wire: BSCCO, known as first generation (1G) wire, and ReBCO, known as second generation (2G) wire. ReBCO stands for "Rare earth - Barium - Copper Oxide" for the superconducting compound. BSCCO stands for "Bismuth - Strontium - Calcium - Copper - Oxygen." Each of these processes has been refined over 20 years time and each type of coated conductor has trade-offs. The driving element that classifies each is operating temperature. Most importantly, by significantly reducing the overall system operating temperature HTS device manufacturers can realize power output increases in the magnitude of 10X.

Medium Temperature Superconducting (MTS) Wire

HTS wire types using a Magnesium di-Boride (MgB) based process are usually produced by reaction of fine Magnesium and Boron powders, thoroughly mixed together and heated at a temperature around or above the melting point of pure Magnesium (> 600 °C).

MgB_2 wires and tapes are therefore realized by means of the so-called Powder-In-Tube method (PIT). MgB_2 systems can be cooled by modern cryocooling devices. The main competing advantages for MgB_2 based HTS wire manufacturing are low cost raw materials and relatively simple deposition techniques.

In contrast, MgB_2's low critical temperature (T_c) of 30 Kelvin is limited to applications that operate at lower temperatures (20K). Low cost continues to be the main driver for MgB_2 wire manufacturers.

However, because of its relatively simple PIT deposition approach, many believe that MgB2 may in the near term better serve applications like electronics in the form of flexible flat ribbon cables and superconducting cavities for RF applications.

MgB_2 is a superconducting wire alternative operating at 20K; a temperature between LTS (4K) and HTS (65K).

- Primary focus for HTS motor and generator applications.

- Price and performance is very attractive.

- Performs very well in high magnetic fields.

- Must operate between 15K to 30K.

- Poor physical properties.

- Cooling costs are more expensive and less reliable than liquid nitrogen.

- Not practical for HTS cable application, although demonstrations are underway.

Low Temperature Superconducting (LTS) Wire

Low Temperature Superconducting (LTS) technology, which operates at liquid helium temperatures (4 Kelvin), was discovered in 1911. This technology became commercially successful in the 1960's when wire was made from LTS materials for use in superconducting electromagnets. LTS electromagnets create fields that are much stronger than conventional copper based electromagnets.

Notably, these state-of the-art LTS electromagnets enabled new technologies like Magnetic Resonance Imaging (MRI) and Nuclear Magnetic Resonance (NMR). LTS superconducting wire is manufactured with Niobium Titanium (Nb-Ti) or Niobium Tin (Nb3Sn) using a powder-in-tube process, embedded in a non-superconducting matrix, such as a silver alloy, somewhat similar to the way traditional wire of copper or aluminum is made. Though LTS wire can be manufactured at costs competitive with copper, LTS devices are very expensive due to the high cost of cryogenic cooling and their reliance on silver. As a result, LTS technology remains quite limited to niche and specialized applications (e.g. Hadron Collider).

In 1987, materials were discovered that exhibited superconducting properties at temperatures as high as 90 K. This class of materials was called High Temperature Superconductors or HTS. While this is still very cold, it was a significant breakthrough. These materials could now be cooled by liquid nitrogen which is much easier to work with, more readily available without supply issues and, most importantly, considerably cheaper than liquid helium.

This drastic cost reduction in cryogenic systems cost opened new opportunities for superconducting applications. HTS communication devices, Maglev transportation, superconducting power cable and superconducting motors and generators were now economically possible. As with LTS devices, many HTS devices used superconducting wire as a base technology.

Features of LTS:

- Are designed only to operate at 4K – therefore limiting to motor, generator applications.

- Cooling cost and reliability key roadblock to market entry – liquid helium required for cooling (scarce resource).

- LTS is in full production use in MRI devices and can be scaled to meet demand for new MRI/NMR devices.

- Good in-field performance and strength – 3T to 10T.

- Very cost competitive - excluding cooling costs.

Yttrium-based 2G Superconducting Wire

Yttrium-based superconducting wire is one of the high-temperature superconductors and is expected to be in practical uses for various applications as the 2nd generation following Bismuth-based superconducting wire.

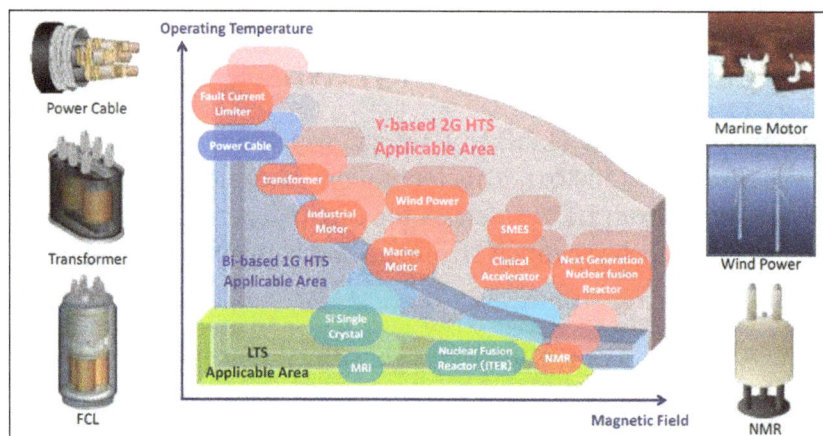

Yttrium-based superconducting wire has higher performance with longer piece length, higher critical current and its uniformity, which will advance various applications to practical use steadily.

The verifications of superconducting applications are ongoing in electrical power equipment such as power cables and fault current limiters, and industrial motors, medical and analytical equipment using superconducting coils. Yttrium-based superconducting wire has higher critical current in magnetic fields than other superconductors and higher critical temperature above liquid nitrogen temperature, which could achieve high-performance superconducting applications at broader ranges of magnetic field and temperature region.

2G HTS Wire offers additional benefits with its unique properties:

- Unmatched critical current capacity.

- Increased in-field performance.

- Significant cost advantages.

- 2G HTS devices are needed today to solve critical challenges in the power grid.

- Increase power capacity.

- Increase efficiency.

- Reduce size, weight and footprint.

- Improve utilization of assets.

Advantages of Superconducting Cable

Compactness and High Capacity

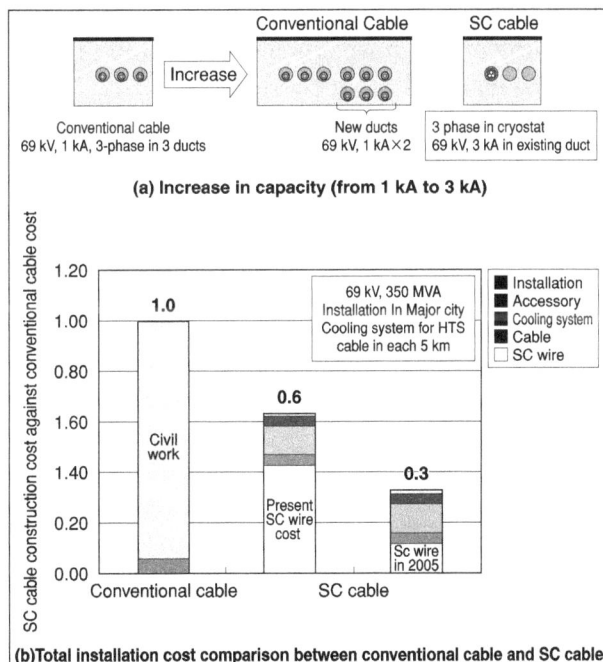

Total installation cost of SC cable.

Superconducting cable can transmit electric power at an effective current density of over 100 A/mm², which is more than 100 times that of copper cable. This allows high-capacity power transmission over the cables with more compact size than conventional cables, which makes it possible to greatly reduce construction costs.

For example, with conventional cable, three conduit lines are normally required to transmit the power for one 66 kV, 1 kA circuit. With these lines, if the demand for electric power expands to the extent that a three-fold increase in transmission capacity is required, six new conduit lines must be installed to lay the new cable. Using a 3-core superconducting cable, however, a 200% increase in transmission capacity can be obtained by installing just one superconducting cable without construction of new conduit. The cost of conduit line construction, especially in a large city like Tokyo or New York, is extremely high. Figure shows a comparison of the construction cost between conventional cable and superconducting cable under the above conditions. The superconducting cable line construction cost, calculated assuming that current construction techniques are used and cooling system is in every 5 km, is likely to be much lower than the cost of constructing a new conduit line of a conventional cable. From the above reason, the superconducting cable is expected to be cost competitive.

Low Transmission Loss and Environmental Friendliness

In superconducting cables, the electrical resistance is zero at temperatures below the critical temperature, so its transmission loss is very small. And the superconducting cable developed by SEI has a superconducting shield, so there is no electromagnetic field leakage outside the cable, which also eliminates eddy current loss from the electromagnetic field. Figure below shows a comparison of the transmission loss in a superconducting cable and a conventional cable. The superconducting cable energy losses come from the alternating current (AC) loss that is comparable to the magnetization loss of the superconductor itself, the dielectric loss of the insulation, and the heat invasion through the thermal insulation pipe. To maintain the superconducting cable at a predetermined temperature, coolant from a cooling unit is required to compensate for this heat gain, and the electric power required for the cooling unit, whose efficiency at liquid nitrogen temperature is thought to be approximately 0.1, must be counted as an energy loss.

Comparing 66 kV, 3 kA, 350 MVA class cables, the loss of the superconducting cable is approximately half that of a conventional cable.

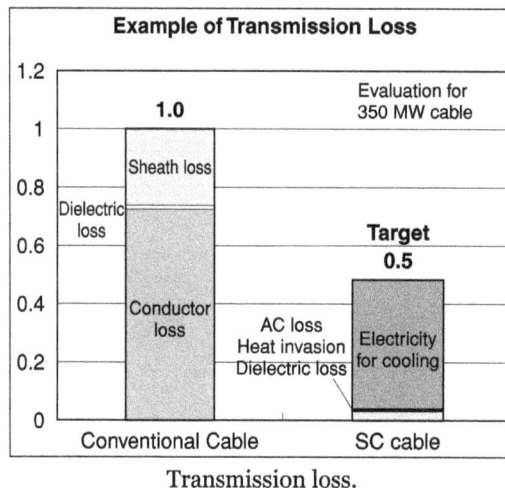

Transmission loss.

Low Impedance and Operatability

A superconducting cable that uses a superconducting shield has no electromagnetic field leakage and low reactance. Depending on the shape of the cable, the reactance can be lowered to approximately one-third that of conventional cables. These features allow the cable line capacity to be increased by laying a new line parallel to conventional cables and controlling the current arrangement with a phase regulator. In this case, it is thought that the control of the phase regulator can be improved and the reliability of the overall system can be increased by using a superconducting cable in the newly added line rather than a conventional cable.

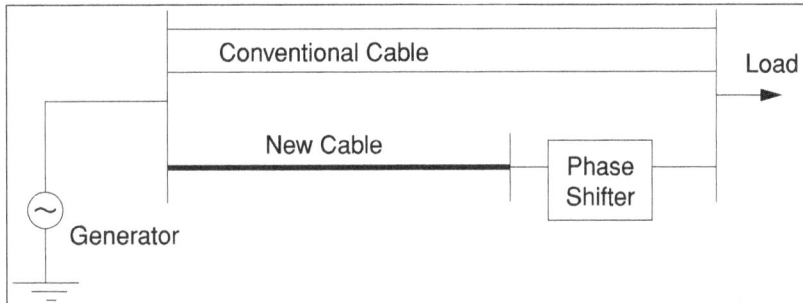

Example of net work with parallel circuit.

In addition, one characteristic of superconducting material is that the lower the operating temperature, the greater the amount of current that can flow. Figure shows the relationship between the superconducting cable temperature and critical current from the SEITEPCO verification test. When the operating temperature was lowered from 77 K to 70 K, there was an approximately 30% increase in the current-carrying capacity. It is hoped that this characteristic can be utilized as an emergency measure when there is a problem with another line.

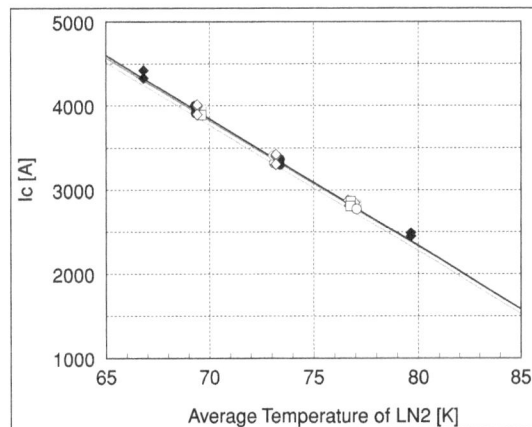

Relationship between cable critical current and temperature in verification tests.

Superconducting Magnets

The development of superconducting magnet science and technology is dependent on higher magnetic field strength and better field quality. The high magnetic field is an exciting cutting-edge technology full of challenges and also essential for many significant discoveries in science and technology, so it is an eternal scientific goal for scientists and engineers. Combined with power-electronic devices and related software, the entire magnet system can be built into various scientific instruments and equipment, which can be found widely applied in scientific research and industry. Magnet technology plays a more and more important role in the progress of science and technology. The ultra-high magnetic field helps us understand the world much better and it is of great significance for the research into the origins of life and disease prevention. Electromagnetic field computation and optimization of natural complex magnet structures pose many challenging problems. The design of modern magnets no longer relies on simple analytical calculations because of the complex structure and harsh requirements. High-level numerical analysis technology has been widely studied and applied in the large-scale magnet system to decide the electromagnetic structure parameters. Since different problems have different properties, such as geometrical features, the field of application, function and material properties, there is no single method to handle all possible cases. Numerical analysis of the electromagnetic field distribution with respect to space and time can be done by solving the Maxwell's equations numerically under predefined initial and boundary conditions combined with all kinds of mathematic optimal technologies.

Magnet Classification

A magnet is a material or object which produces a magnet field. Magnets can be classified as permanent magnets and electromagnets. A permanent magnet is made of magnetic material blocks, has a simple structure and lower costs. However, the magnetic field strength produced by permanent magnets is weak. Electromagnets can operate under steady-state conditions or in a transient (pulse) mode and electromagnets can also be subdivided into resistance and superconducting magnet. A resistance magnet is usually solenoid wound by resistance conductors normally with cooper or aluminum wires and the magnetic field strength is also relatively weaker, but larger than the field generated by permanent magnet. The volume, however, is huge and the magnet system needs a cooling system to transfer the heat generated by the coils' Joule heat. A superconducting magnet is wound by superconducting wires and there is almost no power dissipation due to the zero resistance characteristics of superconductors. The magnetic field strength generated by a superconducting magnet is strong, but limited by the critical parameters of the particular superconducting material. Scientists are trying to improve the performance of superconductors in order to construct superconducting magnets with high critical current density and low operating temperature.

Applied Superconducting Magnet

With the development of superconducting magnets and cryogenic technology, the magnetic field strength of superconducting magnet systems is increasing. A high magnetic field can provide technical support for scientific research, industrial production, medical imaging, electrical power, energy technology etc. Up to now, magnetic fields of about 23 T have been mainly based on low temperature superconductors (LTS), such as NbTi, Nb_3Sn, and/or Al_3Sn. Superconducting magnets with a magnetic field of 35 T are operated in superfluid helium combined with a high temperature superconductor operated at 4.2 K. Magnets with magnetic fields above 40 T are hybrid magnets, consisting of a conventional Bitter magnet and a LTS magnet. Superconducting magnets based on the second generation of YBCO high temperature superconductors may produce a 26.8-35 T magnetic field, while a magnetic field of up to 25 T is possible based on Bi2212 and Bi2223 superconducting magnets. Therefore, research on high magnetic field applications based on superconducting magnet technology has already reached a relatively mature stage.

Magnet in Energy Science

With the global growth of economics and an ever increasing population, energy requirements have been growing fast. Up to now, the available sources of energy around the world are nuclear fission, coal, petroleum, natural gas and various forms of renewable energy. Fusion energy has great potential to replace traditional energy in the future because it is clean and economical. The magnetic field is used to balance the plasma pressure and to confine the plasma. The main magnetic confinement devices are the tokomak, the stellarator and the magnetic mirror, as well as the levitated dipole experiment (LDX). Tokamak has become the most popular thermonuclear fusion device in all countries around the world since the Soviet Tokamak T-3 made a significant breakthrough on the limitation of plasma confined time. The magnetic field strength should be strong enough for the fusion energy to be converted to power and superconducting magnet technology is the best solution to achieve high field strength. The superconducting magnet system of Tokamak consists of Toroidal Field (TF) Coils, Poloidal Field (PF) Coils and Correction Coils (CC). There are several famous large devices including T-3, T-7 and T-15 in Russia, EAST in China, KSTAR in Korea, JT-60SC in Japan, and JET in UK which have been developed and ITER in France will be installed in the future. Figure below illustrates the main technical parameters for the development of some fusion devices.

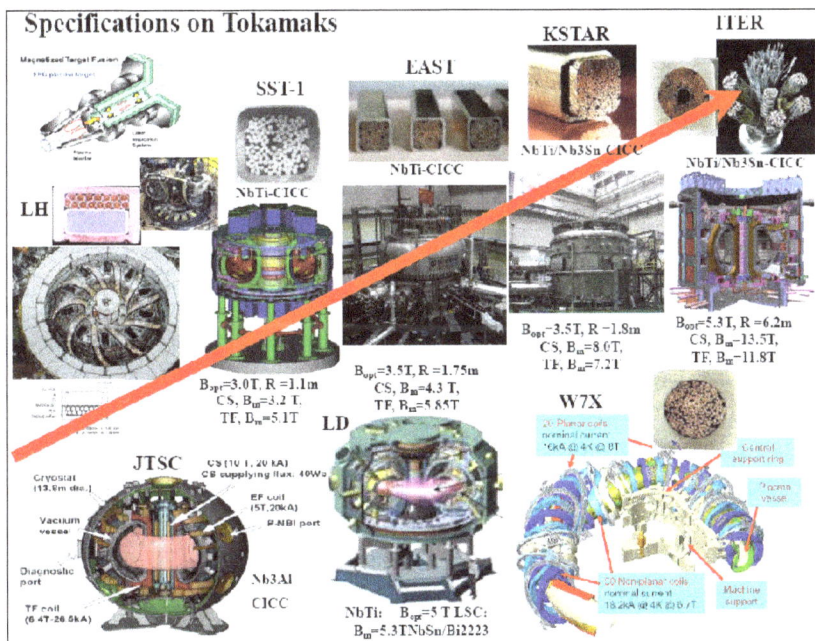

The technical parameters for the development of some fusion devices.

A magnetohydrodynamics (MHD) generator is an approach to coal-fired power generation with significant efficiency and lower emissions than the conventional coal-fired power plant. The MHD-steam combined cycle power plant could increase the efficiency up to 50-60%, which will result in a fuel saving of about 35%. Its applications could provide great potential in improving coal-fired electrical power production. Since the middle of the 1970s, MHD superconducting magnet development has been ongoing and a series of model saddle magnets have been designed, constructed, and tested.

With the commercialization of high temperature superconductors (HTS), various countries and high-tech companies have made great efforts to strengthen their investment in research on superconductivity, and HTS applications have developed rapidly from 1986. At present, HTS cables, current limiters, transformers, and electric motors have already entered the demonstration phase, while experimental prototypes for HTS magnetic energy storage systems have already appeared. Superconducting Magnetic Energy Storage (SMES) technology is needed to improve power quality by preventing and reducing the impact of short-duration power disturbances. In a SMES system, energy is stored within a superconducting magnet that is capable of releasing megawatts of power within a fraction of a cycle to avoid a sudden loss of line power. SMES has branched out from its original application of load leveling to improving power quality for utility, industrial, commercial and other applications. In recent years superconducting SMES systems equipped with HTS have been developed. A HTS magnet with solid nitrogen protection was developed and used for high power SMES in 2007 by IEECAS, and 1 MJ/0.5 MVA HTS SMES was developed and put into operation in a live power grid of 10 kV in late 2006 at a substation in the suburb of Beijing, China. The LTS magnet fabricated with compact structure for 2 MJ SMES consists of 4 parallel solenoids to obtain good electromagnetic compatibility for the special applications. The SMES are shown in figure below.

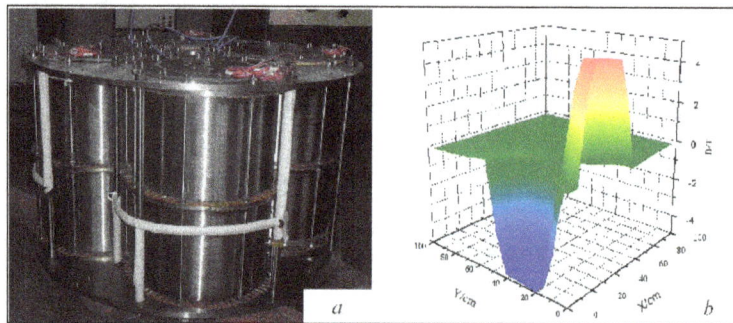

The magnets (a) and (b) The magnetic field distribution 2 MJ SMES

Ultra-high Superconducting Magnet in Condensed Physics

In order to develop a 25-30 T complete high magnetic field superconducting magnet with an HTS magnet system, NHMFL and Oxford Superconductivity Technology (OST) established a collaboration to develop a 5 T high temperature superconducting insert combined with a water-cooled magnet system. They achieved a central field of 25 T in August 2003. By using an YBCO HTS magnet as an insert coil in 2008, the total field was increased to 32.1 T, and a 35.4 T layer-wound YBCO magnet has subsequently been fabricated and tested. The German Institut für Technische Physik (ITEP) at the Karlsruhe Institute for Technology (KIT) used Bi2223 to successfully develop a 5 T insert coil, which operates under a 20 T background magnetic field. The development of this technology provided the technological basis for the development of a high field NMR system. The low temperature required to operate a 20 K HTS magnet can be obtained through a Gifford-McMahon (GM) refrigerator. Because the specific heat at 20 K increases about by two orders of magnitude compared with that at 4.2 K, HTS magnets have higher stability compared with LTS. The HTS magnets with fields of 3.25 T were developed and operated as insert coils in a 8-10 T/100 mm split-pair system in China, the configuration is shown in figure below. The largest HTS magnet project in that laboratory is focused on developing a 1 GHz insert coil. Although the field threshold of Bi2223 and Bi2212 HTS tapes is over 30 T, operation with HTS tapes is limited due to the Lorentz forces. In order to obtain stable HTS magnets, the persistent current mode is used for HTS inserts, with the aim of obtaining field stability smaller than 10^{-8}/h and field uniformity below 10^{-9} in the region of $\Phi10$ mm × 20 mm. The solenoid-type configuration has more advantages than the double pancake structure.

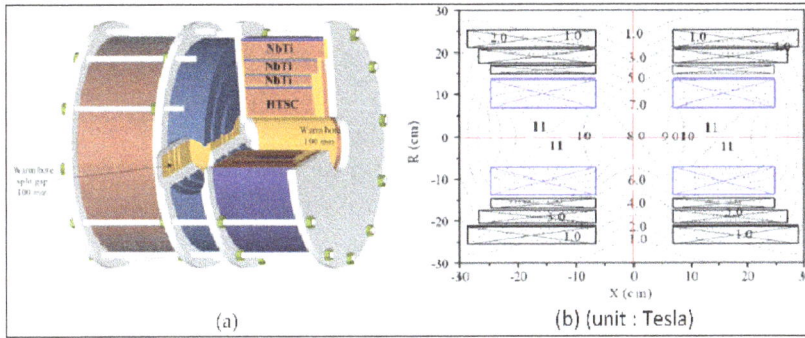

Configuration of 8-10 T/100 mm split-pair (a) and (b) The field distribution.

The 40 T hybrid magnet system will be designed and constructed at the High Magnet Field Laboratory, Chinese Academy of Sciences (HMFLCAS), and the construction of the hybrid magnet is planned to be completed in 2013. The hybrid magnet consists of a resistive insert providing 29 T and a superconducting coil providing 11 T on the axis over a 32 mm bore. The outsert with 580 mm room temperature bore consists of two sub-coils, the inner one (coil C) is a layer wound of Nb_3Sn conductor and the outer one (coil D) is a layer wound of NbTi conductor. Both conductors adopt a cable-in-conduit conductor and will be cooled by 4.5 K force-flowed supercritical helium. For the future upgrade, two Nb_3Sn sub-coils (coil A and coil B) will be inserted into the 11 T superconducting outsert coils and the maximum field in the superconducting magnet will be more than 14 T. Moreover, the resistive insert will be upgraded to 31 T and the total system central field will be above 45 T. Figure below shows the overall configuration and a cross-section of the outsert of the 40 T hybrid magnet system.

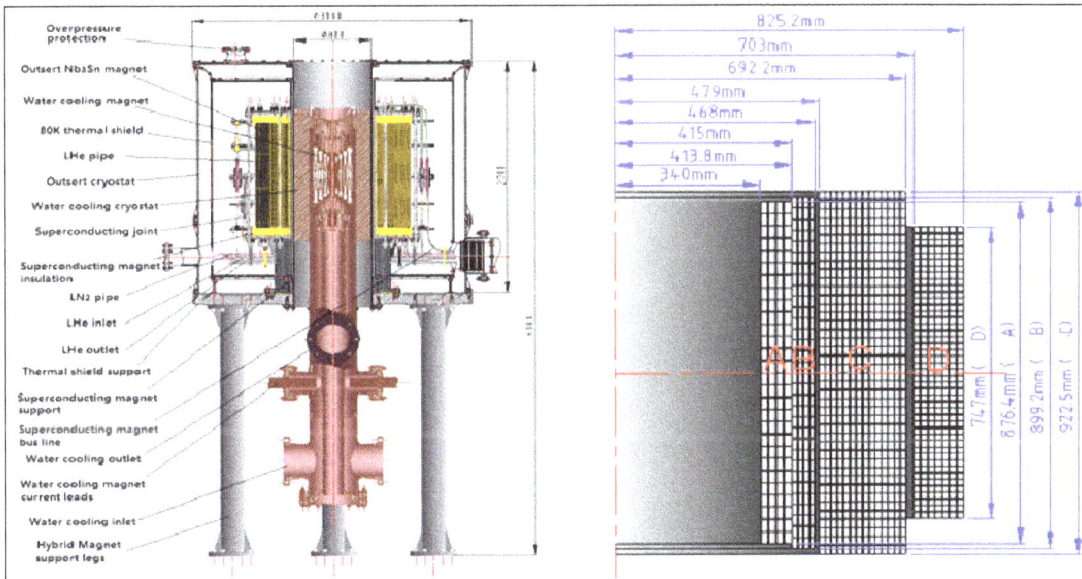

The overall configuration and cross-section of superconducting
outsert of the 40 T hybrid magnet system at HMFLCAS.

Magnet in NMR, MRI and MSS

Since the first Nuclear Magnetic Resonance (NMR) spectrometer magnet system was invented in 1950, NMR has been widely used in leading laboratories all over the world as an effective tool for materials research and it has become the most important analysis tool for modern biomedicine, chemistry and materials science. The use of a superconducting coil for the NMR system (instead of a resistive one) has the advantages of low energy consumption, compact coil structure, stable current and magnetic field, good field uniformity, and high magnetic field. Appropriate super-conductors for high field application are now Nb_3Sn or the ternary compound $(NbTa)_3Sn$. HTS materials, such as YBCO and Bi2212, will be the main superconductors in the future. At present, the standard NMR magnet has an aperture of 52 mm and the magnetic field range is from 4.7 T to 23.5 T. The corresponding frequency is between 200 and 1000 MHz, and the stored energy ranges from 18 kJ to 26 MJ. High field NMR systems need field stability better than 10^{-8}/h and a magnetic field uniformity of 2×10^{-10} in a 0.2 cm³ spherical volume. In 2010, the Bruker Corporation developed a 1000 MHz LTS NMR spectrometer, demonstrating that the LTS conductors NbTi and Nb_3Sn have reached their limit. A 400MHz NMR superconducting magnet system was designed, fabricated and tested at IEECAS. To meet the requirements of 400 MHz high magnetic field nuclear magnetic resonance, the superconducting magnets are fabricated with 17 coils with various diameters of superconducting wire to improve the performance and reduce the weight of the magnet. In order to reduce the liquid helium evaporation, a two-stage 4 K pulse tube refrigerator is employed. The superconducting magnet with available bore of Φ54 mm is shown in figure below.

Configuration of 400 MHz
superconducting magnet with cryostat.

Since 1980, magnetic resonance imaging system (MRI) magnet technology has made continuous progress in medical diagnosis. In the past 30 years, MRI has developed into one of the most important medical diagnosis tools. Due to the clear soft tissue imaging, MRI technology maintains its leading status in medical applications. The key issue in designing and constructing a MRI superconducting magnet is obtaining a highly uniform and stable magnetic field over an imaging volume. The trend in MRI development, therefore, is toward short length of coils, high magnetic field and a fully open, rather than tunnel-like, structure. The shortest coil length up to now is 1.25 m to reduce the patient's incarceration sickness and achieve lower helium consumption.

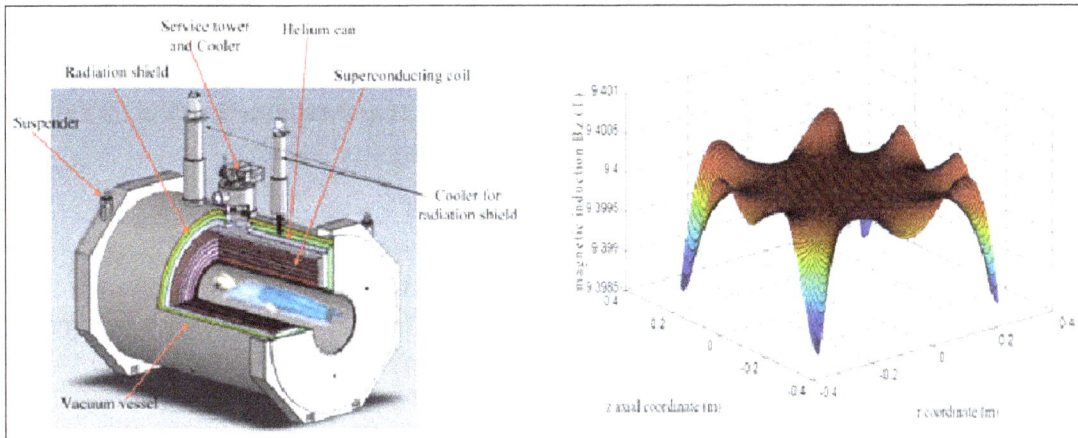

Configuration of cryostat and the field distribution over the DSV region.

At present, the designs of open-style MRI systems use permanent (field range from 0.35 T to 0.5 T) or superconducting magnets. Magnets with fields below 0.7 T can use the combination of a superconducting coil and an iron yoke, which produces a highly uniform field. Standard clinical 1.5 T and 3 T MRI scanners have developed rapidly and now installed in many hospitals. The higher filed devices, may be 7 T MRI, will become the next generation clinical scanner and are supported by three big commercial companies (GE, Philips, and Siemens). The first 7 T whole body scanner with passive shielding was installed in 1999. The first actively shielding 7 T device was designed by Varian and Bruker will soon launch a similar one. The first 9.4 T, which is equivalent to 400MHz functional MRI, was manufactured in 2003 by Magnex Scientific Ltd, a company which was incorporated into Varian. An 11.75 T/900 mm superconducting magnet system is in the process of being fabricated in France; it will be used in neuroscience research in the Commissariat à l'Ènergie Atomique (CEA) in France. Since 2011, a 9.4 T superconducting magnet for metabolic imaging has been undergoing development in the Institute of Electrical Engineering, Chinese Academy of Sciences (IEECAS). The magnet has a warm bore that is 800 mm in diameter and cryogenics with zero boil-off of liquid helium will be used for cooling the superconductors. The overall configuration and the field distribution over the DSV region are shown in figure, respectively.

A magnetic surgery system (MSS) is a unique medical device designed to deliver drugs and other therapies directly into deep brain tissues. This approach uses superconducting coils to manipulate a small permanent magnet pellet attached to a catheter through the brain tissues. The movement of the small pellet is controlled by a remote computer and displayed on a fluoroscopic imaging system. The magnets of the previous generations were composed of three pairs of orthogonal superconducting solenoid coils. The control strategies are complex because of the magnetic field distribution of solenoids. A novel type of spherical coils can generate linear gradient field over a large spherical volume. This type of modified spherical coils with a constant current distribution model is easy to fabricate in engineering. A prototype of this spherical magnet has already been constructed with copper conductors. According to the key research problems of MSS and the disadvantages of the current MSS, we present a novel type of superconducting magnets structure. The first domestic model MSS has also been constructed and a series of experiments have been performed to simulate the real operation situations on this basis.

Magnet in Scientific Instrument and Industry

Initially, superconducting magnets were used as scientific instruments in laboratory. With the improvement of magnetic field strength and performance, superconducting magnet technology has been applied in many fields such as accelerators, industry and so on.

Accelerators are the most important tools for high energy physics research. The investment costs of the accelerator rings are determined by the ring size and the operating costs by the power consumption of the magnets. Superconducting magnets with high current density and lower costs were widely applied to accelerator fields. There are several large, famous accelerators equipped with superconducting magnets such as Tevatron at Fermilab, Hadron Elektron Ringanlage (HERA) at the Deutsches Elektronen-Synchroton (DESY) and the Large Hadron Collider (LHC) at the European Organization for Nuclear Research (CERN). The accelerator superconducting magnet system includes dipole magnets for particle deflection, quadrupoles for particle focusing and sextupole and octupole magnets for correction purposes. It is difficult to design and construct this magnet system because the field distribution is more complex and all magnets need a high effective current density to get the high field strength. In addition, the required high field quality, which means uniformity in the case of a dipole and exact gradient in the case of a quadrupole, and the required repeatability for the series of magnets operated in a high radiation environment are challenges in the design and construction of these magnets.

The Fermilab Tevatron Proton-Antiproton Collider is the highest energy hadron collider in the world. The superconducting Tevatron dipole magnet has a magnetic length of 6.116 m and a radial mechanical aperture of 0.0381 m. The coil package is assembled with an upper coil and a lower coil each of which has an inner layer of 35 turns and an outer layer of 21 turns. The Rutherford-style cable is composed of 23 strands, 12 coated with ebanol and 11 with Stabrite. Each of these strands has 2050 NbTi filaments with the diameter of about 9 μm. The filament separation to diameter ratio is 0.35 and the ratio of copper to non-cooper is 1.8. The coil package is enclosed in a cylindrical cryostat inserted into a warm iron yoke.

The HERA installed at DESY consists of 650 superconducting main magnets (dipoles and quadrupoles) and approximately the same number of superconducting correcting elements (dipoles, quadrupoles and sextupoles). The system consists of two independent accelerators designed to store 30 GeV electrons and 820 GeV protons, respectively. These magnets formed a continuous cold string through the 6.3 km long HERA tunnel interrupted only by warm sections around the interaction regions. The superconducting dipoles with the central field of 4.68 T and the magnetic length of 8.824 m, and the superconducting quadrupoles with the central gradient field of 91.2 T/m and the magnetic length of 1.861 m are of the cold bore and cold yoke type.

The LHC is a gigantic scientific instrument near Geneva, where it spans the border between Switzerland and France about 100 m underground. It is a particle accelerator used by physicists to study the smallest known particles – the fundamental building blocks of all things. Most of its 27 km underground tunnel was filled with superconducting magnets, mainly 15 m long dipoles and 3 m long quadrupoles. The LHC magnets are operated at the field strength of 8.36 T at an operating temperature of 1.9 K, which is approaching the 11.45 T mark that is considered to be the upper limit for a niobium-titanium superconductor. In the LHC accelerator, the stronger the magnetic field is, the tighter the arc of the beam is in its 27 km tunnel. With stronger dipole magnets, an accelerator can push particles to much higher relativistic energies around the same-sized circular beam path.

The Superconducting Solenoid Magnet (SSM) is designed to provide a uniform 1.0 T axial field in a warm volume of 2.75 m diameter. It is the first superconducting magnet of this type built in China. The 0.7 mm diameter NbTi/Cu strands are formed into a Rutherford cable sized 1.26 mm 4.2 mm. The Rutherford cable is embedded in the center of a stabilizer made of high purity (99.998 %, RRR 500) aluminum with outer dimensions of 3.7 mm 20.0 mm. One layer of 0.075 mm thick Upilex-Glassfibre (glass fiber reinforced polyimide) film is used for turn-to-turn insulation of the coil winding. The superconducting magnet is indirectly cooled by a forced flow of two phase helium at an operating temperature of 4.5 K through cooling tubes wound on the outside surface of the support cylinder. The Superconducting IR Magnets (SIM) for the BEPC upgrade (BEPC-II) are installed completely inside the BES-III detector and operated in the detector solenoid field of 1.0 T. In BEPCII, a pair of superconducting quadrupole magnets (SCQ) with high focusing strength is used, which will squeeze the β function at the interaction point and provide a strong and adjustable magnet field. Both of the two SCQs are inserted into the BESIII detector symmetrically with respect to the interaction to produce an axial steady magnetic field of 1.0~1.2 T over the tracking volume and to meet the requirements of particle momentum resolution to particle detectors. figure below shows the components and fabrication at the site of the SCQ.

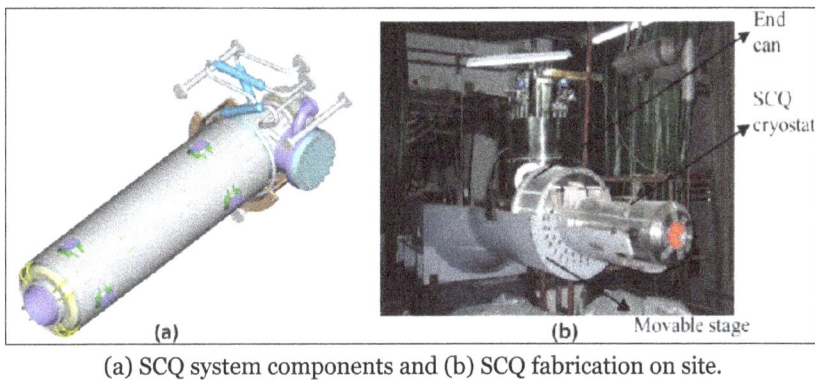

(a) SCQ system components and (b) SCQ fabrication on site.

The MICE coupling magnet consists of a single 285 mm long superconducting solenoid coil developed by the Harbin Institute of Technology. The superconducting coil is wound on a 6061 aluminum mandrel that is fitted into a cryostat vacuum vessel. The inner radius of the coil is 750 mm and its thickness is 110 mm at room temperature. The coil assembly is comprised of the coil with electrical insulation and epoxy, and the coil case is made of 6061-T6-Al, including the mandrel, end plates, banding, and cover plate. The length of the coil case is 329 mm. The coupling solenoid will be powered by a single 300 A/020 V power supply connected to the magnet through a single pair of leads that are designed to carry a maximum current of 250 A. It is cooled by liquid helium flow through cooling tubes embedded in the coil cover plate by two 1.5 W cryocoolers.

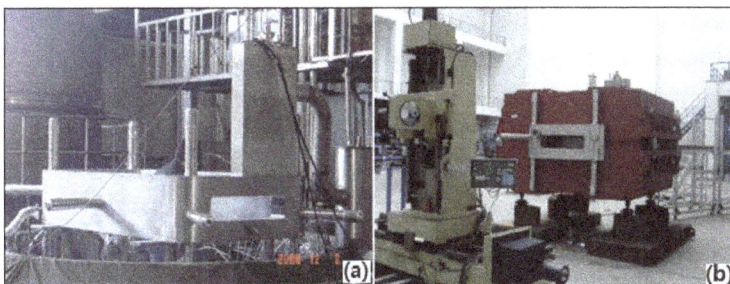

(a) Cryostat and (b) Prototype of dipole magnet for GSI CR ring.

The super-ferric dipole prototype of the Super Fragment Separator (Super-FRS) has a width of 2200 mm, a central length of 2020 mm and a height of 725 mm, respectively. The Collector Ring (CR) magnet has been built by the Facility for Antiproton and Ion Research (FAIR) China Group (FCG), including IMPCAS, IPPCAS, and IEECAS, in cooperation with the GSI Helmholtz Center for Heavy Ion Research (GSI) (Hanno Lei brock et al. 2010). IEECAS is contributing to the design of the coils, IPPCAS is responsible for the fabrication of the superconducting coils and cryostat and IMPCAS is responsible for the testing of the whole system, the magnetic field optimization and the development of the 50 ton laminated iron yoke. The dipole magnet has a homogeneous region 380 mm in width and 140 mm in height, while the homogeneity reaches $\pm 3 \times 10^{-4}$. The passive air slot and chamfered removable end poles guarantee that the magnetic field distribution is homogeneous at both low and high field levels. The Super-FRS superconducting dipole is a super-ferric super-conducting magnet with a warm iron yoke which is laminated due to magnetic field ramping and the H type yoke is made of laminated electrical steel 0.5 mm in thickness, which was stamped and glued to blocks, and machined to the 15 degree angle sector shape. The superconducting coils were wound from multi-filamentary NbTi wires with a higher than usual the ratio of copper to non-copper and are operated at liquid helium temperature. The coils are positioned in the helium cryostat with a multi-layer insulation structure. The total weight of the magnet is more than 52 tons. The magnetic field measurements indicate that the field homogeneity is about $\pm 2 \times 10^{-4}$ at different magnetic field levels (0.16 T - 1.6 T), which is better than the design requirements. The cryostat at the test facility in IPPCAS and prototype dipole at the test facility of IMPCAS are shown in figure.

For the requirements of microwave devices, a conduction-cooled magnet has been fabricated for the microwave experiments used in Gyrotron. A magnet system with a center field of 1.39 T and warm bore of 100 mm has been designed and fabricated. The electromagnetic structure of the magnet is designed on the basis of the hybrid genetic optimal method. The length of homogeneous region of the superconducting magnet is adjustable from 200 mm to 250 mm. Also the superconducting magnet can generate multi-homogeneous regions with the length of 200, 250 and 320 mm. The homogeneity of the magnetic field is about 0.5% with a constant homogenous length and 1.0% for adjusting homogenous length. All of the homogeneous regions start at the same point and the field decays to 1/15-1/20 from the front point of a homogeneous region to 200 mm. The superconducting magnet is cooled by one GM refrigerator with cooling power of 1.5 W at 4 K. The configuration of the superconducting magnet with superconducting coils and copper coils is illustrated in Fig. The homogeneous region length of 320 mm with maximum center field of 1.3T is generated by main coils 1, 2 and compensating coils 3 and 5. In order to ex-tend the magnetic field decay from 200 mm to 300 mm, we need to use the normal copper coils. Therefore, the homogeneous region length of 320mm with a field of 1.3 T is generated by the main and compensating coils of 1, 2, 3, 5, 6 and 7 for superconducting coils and 8, 9 and 10 for copper coils. The homogeneous region length of 250 mm with the field of 4 T, and the magnetic field decay can be adjustable through the main and compensating coils of 1, 2, 3, 5, 6 and 7. The magnetic field decay can be controlled through the adding cathode compensation coils of 6 and 7, and the coils are connected to an assisting power supply to adjust the operating current. The total superconducting coil set-up should have five high temperature superconducting current leads. The copper adjustment coils 8, 9, and 10 are used to change the operating current to cor-rect the magnetic field distribution in the homogeneous region. The main field distributions are illustrated in figure. For the requirements of IEECAS customers, superconducting magnets with all kinds of magnetic field distribution are fabricated.

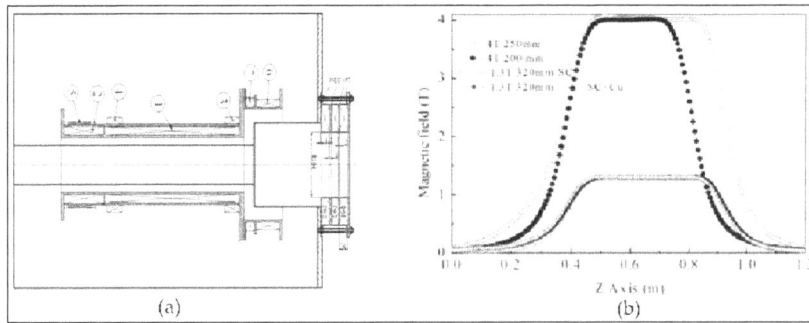

(a)   (b)

Superconducting magnet with 10 coils arranged with the same axis: the superconducting coils are installed in the cryostat; the copper coils are located outside of the cryostat and fixed on its flange, where they are cooled by air convection. The superconducting coils are cooled by a GM cryocooler (a), Magnetic field distribution for various lengths of homogeneous region (b).

Structure and Function of Magnets

The desired magnetic field produced by superconducting coils and the shape of field is predetermined by the users and its special application. The magnetic field distribution depends on the size and shape of coils and final system structure. The common shapes of superconducting coils are solenoid, saddle coils, race-track coils, toroid coils, baseball coils and yin-yang coils for different applications.

Configuration of Solenoid Magnet

The most efficient and economic coil is the solenoid structure, and normal solenoids are symmetric consisting of a single solenoid or several coaxial solenoids based on the field distribution and homogeneity demands. The solenoid coil is wound layer by layer with round or rectangular cross-section wires on a cylindrical bobbin. The basic parameters for a solenoid are inner radius r_{inner}, router radius r_{outer}, the length L and the current density J. The current density $J = NI_{op}/[L(r_{outer}-r_{inner})]$, where the number of windings and operating current are N and I_{op}, respectively. The conductor current density is higher due to the electrical insulation and the eventual mechanical reinforcement. By these parameters, the magnetic field can be calculated by the popular equation. By this method, in theory, we can design all symmetric field distribution magnet system.

Racetrack and Saddle-shaped Magnet

The racetrack-shaped coil has two linear segments and two semicircular arc segments. The saddle coil has two linear segments and six small circular segments. The coil structure of racetrack-shaped and saddle coils are shown in Fig. The racetrack-shaped magnet may be used in electrical machinery, magnetic levitation trains, dipole or multipole magnets for an accelerator, wiggler and undulator magnets, large-scale MHD magnets, space detector magnets and in space astronaut radiation shield and accelerator detector magnets, such as the ATLAS magnet at CERN. Sometimes, accelerator magnets, electrical machinery magnets and MHD magnets employ saddle shaped coils. Transverse magnetic field distribution can be produced by combining with the saddle coils and change the current direction. Saddle coils are also used to correcting the magnetic field distribution for magnetic resonance magnets and magnetohydrodynamics.

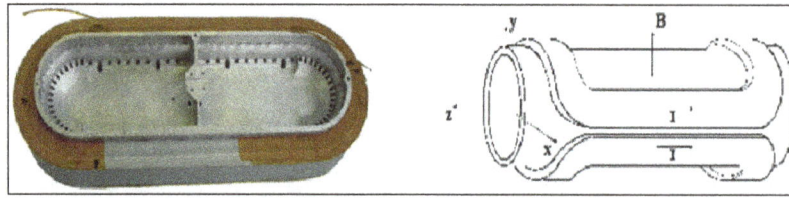

Configurations of a racetrack magnet and a saddle magnet.

Structure of other Complicated Magnet

Baseball coils and yin-yang coils is used to confine the plasma as magnetic mirror. The baseball coils with U-shaped structure produce a magnetic field magnitude increasing in every direction outwards from the plasma and the structure is more economic than the same mirror field produced by a pair of solenoid. A yin-yang magnet consists of two orthogonal baseball coils which generally produce a deeper magnetic well than a single baseball coils and also use fewer conductors. The magnet structure of a magnetic mirror is even more complex, as for the stellarator shown in Fig. The magnet current distribution forms a yin-yang structure. Force-free magnets are ones in which the current density J is parallel to the field H everywhere, i.e. $J = H$, where is a scale function called the force-free function or factor. The Lorentz force f is therefore equal to zero since f = μJ H = o. However, from the virtual work theorem of mechanics, it can be verified that it is impossible to be force-free everywhere in a finite electromagnetic system without magnetic coupling with other systems. Furthermore, it is also unnecessary to construct a fully force-free magnet, as shown in figure since the solid coil itself could withstand certain forces. So we need practically to develop some quasi-force-free magnets in which J and H are approximately parallel, so that although the Lorentz forces are not zero, they are reduced significantly. With the development of accelerator magnet technology in recent years, the so-called snake-shaped dipole magnet, shown in Fig. has been proposed, which can deliver good magnetic field uniformity. This kind of magnet can be used in accelerators for particle focusing.

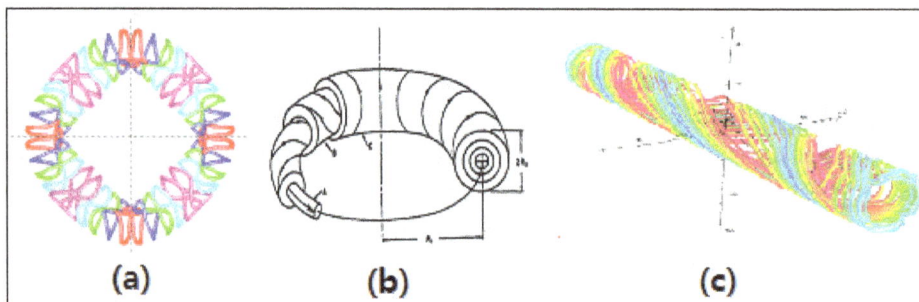

Configuration of (a) stellarator, (b) force-free, and (c) snake-shaped magnets.

Numerical Methods for Magnet Design

Due to the complex structure of electromagnetic devices and the compact design requirements, the design of modern magnets no longer relies on simple analytical calculations. Usually, the designers employ complex high-level numerical analysis technology to decide the electromagnetic structure parameters. With the geometry, the material distribution and the driven sources given, the numerical analysis of the electromagnetic field distribution with respect to space and time can be conducted by solving the Maxwell's equations numerically under predefined initial and boundary conditions. During the design of magnet devices, the designer should propose a configuration satisfying the functional needs as far as possible. The inverse problem is: given the magnetic field distribution in space and time, one must find the geometric parameters and the material distribution, as well as the field source. Magnet design is such an electromagnetic field inverse problem. Its task is to find the field source (current distribution or permanent magnet material distribution) on the basis of a given magnetic field spatial distribution. The inverse problem has two different aspects: the design optimization itself and the parameter identification.

The deterministic method is doing the search gradually during the iterative process according to the search direction determined by each step of the iteration, so that the objective function value of the current step iterative solution is certain to be smaller than the preceding values. Different deterministic methods have different search directions, such as the "steepest decent", the "conjugating gradient method", the "Quasi-Newton law", the "Levenberg-Marquard algorithm", etc. The deterministic method depends on the neighborhood characteristics of the current search position to determine the next step search position (with partial linearization in a non-linear problem). Therefore it is local optimization and the efficiency of seeking for the local optimal solution is very high, but it does not have global optimization capability in a multi-extreme value problem. Another shortcoming of the deterministic method is that it is necessary to know the first- or second-order partial derivative of the objective function and it usually requires the objective function to not be too complex and to have an analytic expression, which increases the computing time and cost. On the other hand, the ill-posed inverse problem is often inherited with an optimization problem. Therefore, the regularization processing should be added to each iteration step of the deterministic method, as otherwise big errors will occur, and the iteration may not work.

In order to avoid the limits of the deterministic method, the stochastic method (Monte Carlo) is suggested. The Monte Carlo method works in such a way that each iteration step is determined by a random number. The traditional Monte Carlo method carries on a completely stochastic blind search, assuming that all possible solutions have equal probability. In contrast, the modern Monte Carlo methods, such as the well-known simulated annealing method, the genetic algorithm, the evolutionary algorithm (ant colony algorithm and particle swarm optimization), the taboo search method, and the neural network and other stochastic algorithms, carry on the random search in a more instructive way, giving the different possible solutions with different probabilities. The merits of the Monte Carlo method are: it is universally serviceable and no target problems need to be differentiated as to whether they are linear or non-linear, ill-posed or well-posed. A problem can be processed by the Monte Carlo method, even if its operator is very complex and cannot be expressed with an analytical formula. Besides, the method has a strong optimization capability in all situations. Its shortcoming is that the calculation time is usually very large, growing inordinately with the order of the problem, while the convergence rate is very slow.

In order to combine the respective merits of the above algorithms, many researchers have been striving to work for the unification of these methods. In order to reduce the computing time, a new kind of optimizing strategy has emerged in recent years – the unification of the response surface model and the stochastic optimized algorithm. This method firstly separates the space of the target variable for a series of sampling points and then applies the numerical calculus method to compute the value of the objective function on these sampling points; with these values, it uses a response model to reconstruct the objective function and then the optimization computation is carried out using the optimizing algorithm on the restructured objective function. Because it is only necessary to calculate the value of the electromagnetic field objective function on the sampling points, the algorithm efficiency is enhanced greatly. Sometimes in the optimization design of an electromagnetic installation, unifying the Moving Least Squares method with the simulated annealing method has very good results. The convergence rate of these algorithms, however, still cannot satisfy the requirements for computing complex large-scale systems, for example, three-dimensional calculations, transient processes or coupled systems, at present. In magnet design, a combination of the deterministic and the stochastic algorithms has been adopted.

Design Example of High Homogeneous Magnet

A high homogeneous magnet system is the most important and expensive component in an MRI or NMR system. A superconducting magnet with the distribution of coils in single layer, two layers and even more layers is the best solution to achieve the high magnetic field strength and homogeneity requirements. The challenge for designing a MRI magnet system is to search the positions and sizes of coils to meet the field strength and homogeneity over the interesting volume and stray field limitation.

Mathematical Model for a Hybrid Optimization Algorithm

The parameters of the system, including length, inner and outer radius of feasible region of coils, radius of DSV region and the axial and radial radius of 5 Gauss stray field line, are predetermined by designer based on the actual applications. Fig. illustrates an example for the design of a symmetric solenoid magnet system. The required parameters of the feasible rectangular region for arrange coils are inner and outer radius (r_{min}, r_{max}) and the length (L), the interesting imaging volume is commonly sphere and the stray field is limited to smaller than 5 Gauss outside the scope of an ellipse. A hybrid optimization algorithm by combination of Linear Programming (LP) and Nonlinear Programming (NLP) was developed by IEECAS. This approach is very flexible and efficient for designing any symmetric and asymmetric solenoid magnet system with any filed distribution over any shape volume.

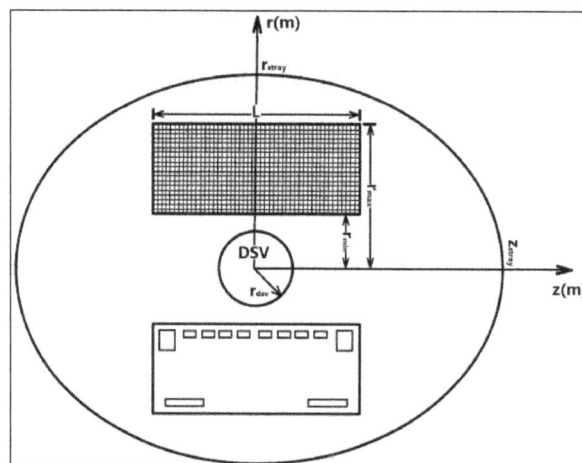

The region of feasible coils, interesting volume and stray field.

Firstly, the feasible rectangular region can be meshed as 2-D continuous elements with N_r elements for radial direction and N_z elements for axial direction, and each element served as an ideal current loop. The surface of the sphere and the ellipse for homogenous filed and stray field limitation were evenly divided into N_d and N_s parts along the elevator from 0 to π, respectively. The field distribution at all target points including N_d and N_s points produced by all ideal current loops with unit current amplitude calculated, and the unit current field contribution matrices A_d and A_s were formed. A LP algorithm was built up with the objective functions totaling the volume of superconducting wires, the field distributions at all target points and the maximum current aptitude for all current elements were constrained. The LP mathematical model is formulated as following:

$$\text{Minimum: } \sum_{i=1}^{Nz+Nr} r_i |I_i|$$

$$\text{Subject to: } \begin{cases} |A_d * I - B_0| / B_0 \le \varepsilon \\ |A_s * I| \le 5 \; Gauss \\ I \le I\max \end{cases}$$

The current map with sparse nonzero clusters were calculated by first LP stage, the positions of nonzero clusters can be discretized into several solenoids with the size of inner radius (r_{inner}), outer radius (r_{outer}), and the z position of two ends (z_{left}, z_{right}). Secondly, a NLP algorithm was built up, and the objective function and constraints of the algorithm are similar to the LP stage, and added current margin constraint into the algorithm which based on the maximum magnetic field within the superconducting coils and the relationship of critical current and magnetic field. The NLP mathematical model is formulated as following:

$$\text{Minimum: } \sum_{i=1}^{Ncoils} V_i$$

$$\text{Subject to: } \begin{cases} \left[\max\left(B_{zdsv}\right) - \min\left(B_{zdsv}\right) \right] / mean\left(B_{zdsv}\right) \le H \\ \sqrt{B_{zstray}^2 + B_{rstray}^2} \le 5 Gauss \\ I / Ic\left(B_{\max}\right) \le \eta \end{cases}$$

here, the Ncoils is the number of discretized solenoids, V_i is the volume of the ith solenoid, B_{zdsv}, B_{zstray} and B_{rstray} are the magnetic field on DSV region and stray field ellipse region, respectively. η is the current margin, B_o is the target field over the DSV and H is the homogeneity level for design.

Many cases were studied by this hybrid algorithm, and the results show the method is flexible and efficient for the first LP stage which took about 5 minutes and the second NLP stage which took about 30 minutes, respectively. The resultant coil distributions were simple and easy to fabricate.

Design Cases

Actively Shielded Symmetric Solenoid MRI

The actively shielded symmetric MRI system with the length of 1.15 m, the central magnetic field of 1.5 T and the field quality of 10 ppm over 500 mm DSV and the radial inner and outer radius of 0.40m and 0.80 m, the stray field of 5 Gauss outside the scope of an ellipse with axial radius of 5 m and radial radius of 4 m, and the current margin of 0.8. The N_r, N_z, N_d, and N_s were set as 40, 40, 51 and 51, respectively. The current map by LP and the final actual coils sizes and positions are shown in figure. The coils distribution with two layers, the inner layer with four pairs of positive and one pair of negative current direction coils for producing the required magnetic strength and the homogeneity. The outer layer with a pair of negative current direction coils for reduces the stray field strength. The homogeneities and stray field distributions are shown in Fig. The coils parameters are shown in Table I. The operating current density and actual current margin are 148 MA/m² and 0.7546, the maximum magnetic field and hoop stress are 5.43 T and 145.16 MPa, respectively.

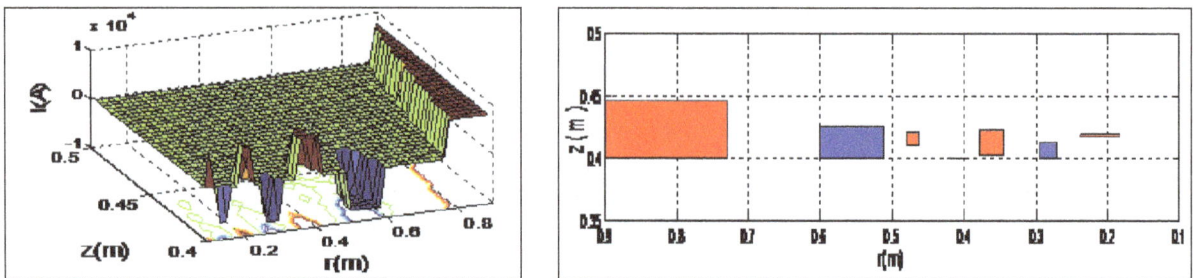

The current map and coils distribution of half model.

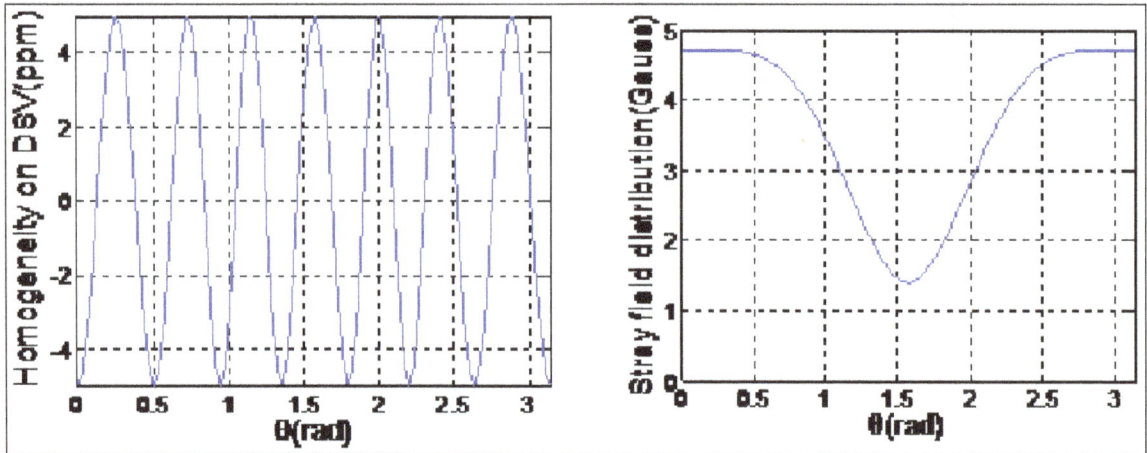

The homogeneity over 500 mm DSV and stray field distribution.

Open Biplanar MRI

The unshielded open biplanar MRI system has a central field strength of 1.0 T, inner and outer radius of 0.0 and 0.90m, lower and upper z positions for two ends of 0.40m and 0.45 m, and field quality of 15ppm over 450 mm DSV.

The N_r, N_z, N_d, and N_s were set as the same as the design of 1.5 T actively shielded MRI system. A quarter model current map for the LP stage and the coils distribution are shown in Fig. The coils distributions with single layer have six coils, the largest coils with maximum field of 5.26 T are the outermost coils with positive current direction. The operating current density and actual current margin are 150 MA/m² and 0.74, the hoop stress is 128.2 MPa, respectively.

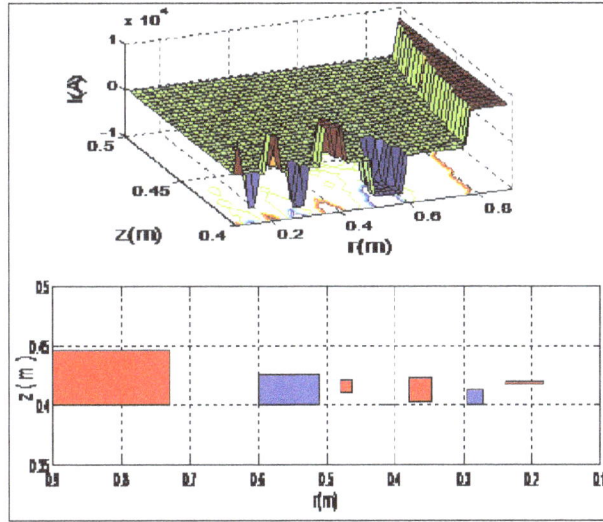

A quarter model current map and coils distribution.

Superconductive Fault Current Limiters (SFCL)

All parts of a power system have to be designed with proper considerations to be capable of withstanding the thermal and mechanical stresses caused by short circuits or "fault" currents.

These fault currents may exceed the nominal current by the factor of 100. Therefore reduction of the fault current would contribute to a significant cost and energy saving as well as optimization in final design.

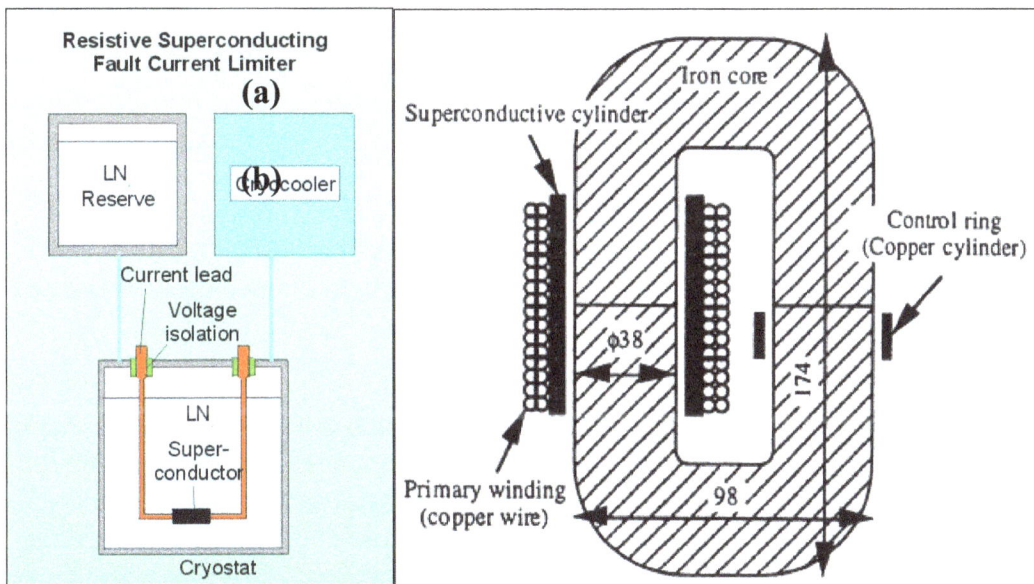

Superconductive FCLs (a) resistive type (b) shielded core type.

Various fault current limiters (FCL) are designed and developed to overcome the fault current problems in power systems. Among these various types, superconductive fault current limiters (SFCL) offer the most ideal performance since in normal operation they are in superconductive state and have no impedance. This is only during the short circuit that they quench into normal state as a result of their critical current being surpassed and enter high limiting impedance into the circuit. This no-impedance operation in normal state and high-impedance operation in fault state is only offered by the superconductive FCLs.

There are two major types of SFCLS namely resistive and shielded core; the former being characterized by an inductance when the fault occurs while the latter simply enters a resistant in the circuit to limit the fault current.

SFCL allows for novel design of power systems; techniques to enhance the reliability of power systems such as coupling of grids and paralleling can be conveniently used with no concern for the increment in fault current level of the system. Because of the advanced level of the SCFCL technology, they are expected to be the first large-scale applications of HTS.

Superconductive Transformers

Superconductive transformers are some of the other successful applications of superconductors in power industry.

Superconductive transformers have more current density than conventional ones and less copper losses. This makes applicable the power transformation with more efficiency and less volume. Moreover it is possible to use more wire turns in these transformers corresponding to less "V/T" – Volt per Turn – to reduce the core dimensions. With respect to the fact that size and weight of the transformers are mostly dependent upon their core, this reduced core dimensions would in turn contribute to great reduction in the final size and weight of the device.

Superconductive transformers usually have less impedance that improves the dynamic of the system. Since they are very similar in construction to the SFCLs they can limit the fault currents as well. This would prevent later damages to other devices and the transformer itself.

Another attractive feature of Superconductive transformers is their capability to work oil free. This is valuable in places that transformers are subjected to fire hazards which is an ordinary case in chemical and petroleum plants. Other advantages of these transformers are: capability to work continuously in overload conditions without any lifetime loss –because of the ultra-cold operating environment-, improvement in volt-age regulation, possibility to remove the core and etc.

Superconductive Motors and Generators

Two new types of HTS electric machines are considered. The first is hysteresis rotors containing bulk HTS elements. The Second is reluctance motors with component HTS-Ferromagnetic rotors, consisting of joined alternating bulk HTS (YBCO) and ferromagnetic (iron) plates, provide a new active material for electromechanical purpose. Such rotors have anisotropic properties (ferromagnetic in one).

Magnetic fields in conventional motors and generators are created by large coils of copper or aluminum wire. HTS wires have much higher current capacities, which means considerably smaller and more powerful motors and generators can be built. In addition, due to the much lower value of electrical resistance in superconductive machines, they have higher efficiencies compared with conventional copper machines.

Figure below is cross section view of the generator build at Emerson Electric. The rotor has 8 iron posts. The stator of the generator was wound with copper wire in a 3 phase, 8 pole configurations. The generator was submerged in a container of liquid nitrogen, and was driven by an external motor mounted above the generator. The generator was loaded with three phase variable resistance bank. The bearing system consisted of conventional open ball bearing that had been cleaned of all grease. The liquid nitrogen acted as the bearing lubricant.

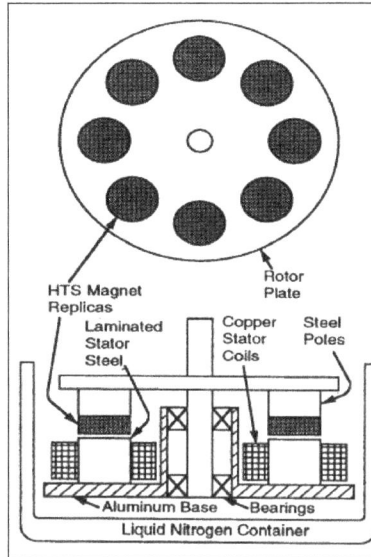

Cross section of Emerson Electric/TCSUH generator. Top: View of rotor, with HTS magnet replicas mounted to the steel poles. Bottom: Side view of the rotor, stator, and cryostat.

Estimating Future Markets for High Temperature Superconducting Power Devices

In the distribution of losses in the national grid is carefully traced and those losses that HTS can eliminate are identified .The energy savings achievable by the many sizes of HTS generators, transformers, cables, and motors are then computed and totalled using a spreadsheet analysis. The economic savings are very sensitive to the price (and) of HTS wire, and to the cost of cooling the devices to operating temperature. A market penetration model is used to estimate how fast HTS devices become commercially successful. The result is shown in figure below.

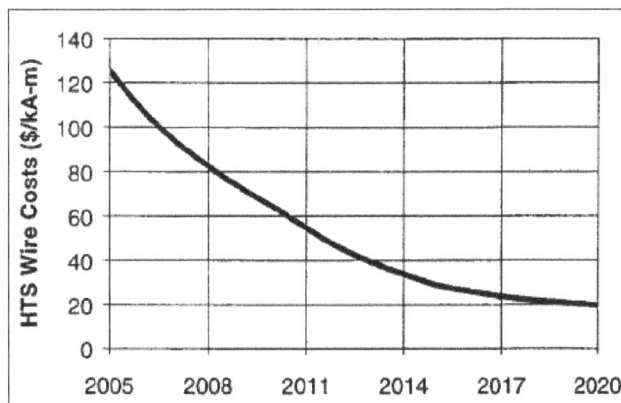

Cost trajectory over time for HTS wire (in dollars per kiloAmperes per meter.

Embedded in this national spreadsheet analysis are models of the behavior of the four main HTS devices (motors, generators, transformers, and cables) that would be sold to utilities and industrial customers.

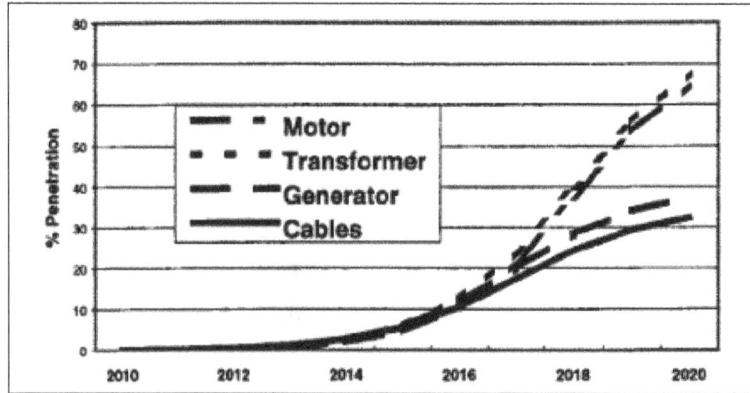

Market penetration curves for HTS motors, transformers, generators, and cables.

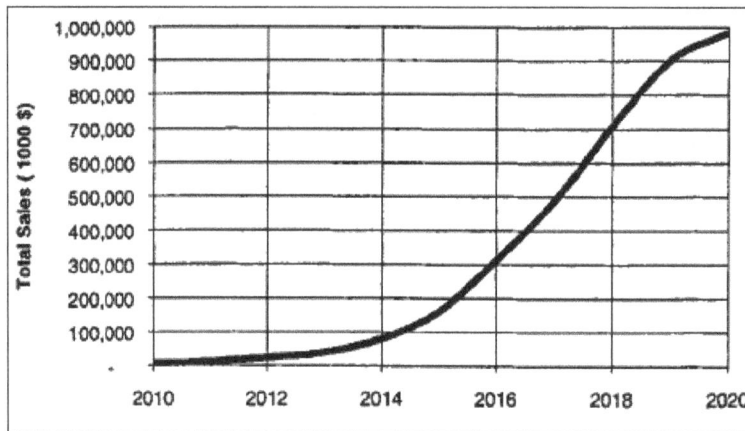

Estimated total sales for all four categories of HTS devices over time that result from the combination of input values used for market penetration, wire, and cryogenic parameters.

Application in Automobiles

The great majority of land-transport machines (mainly automobiles) consume large quantities of petroleum, either in the form of gasoline or diesel fuel. To cope with the issues of depletion of fossil fuels and global warming caused by CO_2 emissions and realize sustainable motorization, significant improvements in energy efficiency and shift to non-petroleum alternative fuels are required. There are mainly three types of environmentally-friendly vehicles under development by automakers and research institutes: Hybrid electric vehicle (HEV), electric vehicle (EV), and fuel cell vehicle (FCV).

HEVs are under mass production since 1997, spreading much faster than EVs or FCVs. Because HEVs use a combination of a gasoline engine and an electric motor, the existing infrastructure for gasoline-driven vehicles can be utilized. EVs and FCVs, meanwhile, produce zero emissions during operation, making them the ultimate eco-cars. EVs offer strong overall improvements in energy use and environmental impact, as electricity is available with relatively low petroleum use. Instead, EVs rely on low CO_2 emitting energy sources such as nuclear power, gas power, and hydraulic power. FCVs likewise provide substantial reduction of CO_2 emissions, as hydrogen used as a fuel can be produced from various clean energy sources. FCVs are expected to spread rapidly and replace the conventional internal combustion engine vehicles in the future, once improvement is seen in terms of battery capacity, cost, and infrastructure installation.

What these three types of environmentally-friendly vehicles have in common is the use of electricity to deliver power from a motor. Realizing high efficiency, high performance motor is the key to the development of electric vehicle systems. Thanks to recent improvements in the performance of high-temperature superconducting wire and in the adiabatic cooling technology, the feasibility of more efficient electric vehicle system by applying superconducting motor is being actively investigated.

Figure below shows the features of a superconducting motor coil. A superconducting coil provides a high magnetic flux density, and therefore delivers much higher torque than ordinary motors. Also, a superconducting motor can be used without copper loss, and an air-core superconducting motor may be developed in the future to reduce iron loss and increase motor efficiency. The drive range of an automobile motor is wide, from low to high speeds and from low torque for constant-speed cruising to high torque for acceleration. While an ordinary motor exhibits increased copper loss and poor efficiency during high-torque output, a superconductive motor provides high efficiency over a wide range.

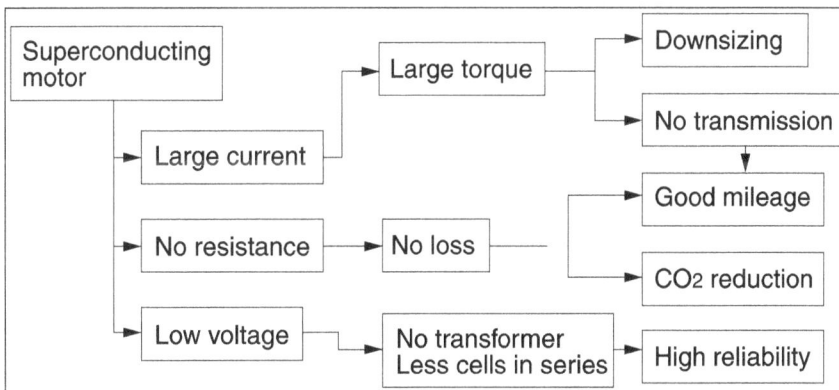

Features of superconductor motor.

The structure of a system for applying superconductivity to an ordinary electric car engine is shown in figure. With an ordinary motor, drive power is supplied in different revolution speed ranges, from several thousand to more than ten thousand RPM, and variable-speed gears are used to reduce the revolution speed and increase the torque. A superconducting motor, on the other hand, would supply much higher torque, so the system can be designed with the motor directly driving a shaft, without variable-speed gears. This should allow the reduction of transmission loss caused by the gear system.

The challenge in applying a superconducting motor to an automobile is that the superconducting coil must be kept at a low temperature, below the critical temperature of the superconducting material. A refrigerating mechanism is required, as it is necessary for the coil to be at a low temperature at the time it is utilized, and it is necessary to keep it at a low temperature using a refrigerating unit while the vehicle is in operation. Therefore, it is most advantageous to use such a system in vehicles that are utilized at a high rate of operation.

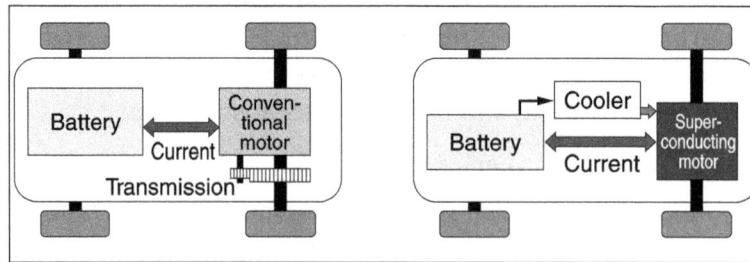

System configuration.

Also, in a heavy vehicle that requires a high output, the output of the cooling mechanism has less impact on the increase of motor efficiency and the decrease of transmission loss. Furthermore, because high output is required during acceleration or deceleration regeneration of a heavy vehicle, superconducting motor would also be suitable for buses and other mass-transit vehicles that experience frequent stop-and-go operations.

Prototype Superconductive Motor

Principle, Wire and Specifications

The specifications of the tested motor are shown in table. In terms of basic principle, it is a series-wound DC motor with a coil formed of the polyimide-film-wrapped type-H superconducting wire to provide a constant field. The superconductive coil is immersed in liquid nitrogen for refrigeration, and a stainless-steel cooling insulation vessel with a vacuum layer is used. The iron core consists of four claw poles. The coil is a simple pancake coil formed of 186 turns of superconducting tape. The claw pole design was adopted to enable larger coil winding radius and thus reduce the number of coils, and to simplify the design of the cooling apparatus and other elements. Because the motor has a long shaft, the coil is divided into two parts as shown in figure. The same armature rotor as that used in a commercial DC motor is used.

Table: Specifications.

Wire	Wire type	Type H (polyimide film insulated)
	Dimensions	4.2 × 0.22 mm
	Critical current (Ic)	140A
	Maximum tensile strength (77K)	150MPa
	Minimum bend radius (room temperature)	70mm
Coil	Shape	Inner dia.: 186 mm Outer dia.: 210 mm Width: 40 mm
	Turns	186 turns / coil
Motor	Motor type	Series-wound DC (field superconductor
	Coil refrigeration method	Liquid nitrogen immersion
	Maximum voltage	144V
	Maximum current	500A
	Dimensions	267 (dia.) × 370 (height) mm
	Weight	About 70 kg

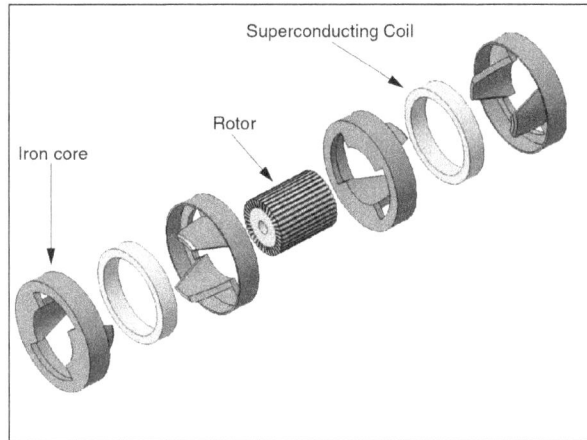

Component s of superconducting motor.

Design

In designing a motor using a superconductive coil, the configuration of claw poles was determined by means of CAE analysis of magnetic flux density distribution and magnetic saturation. Figure is an example of an analysis image. In figure, the measured torque values are plotted over the designed current/torque characteristics of the motor. The measured drive torque of the motor mounted on a vehicle was very close to the design value of 58 Nm at a maximum current.

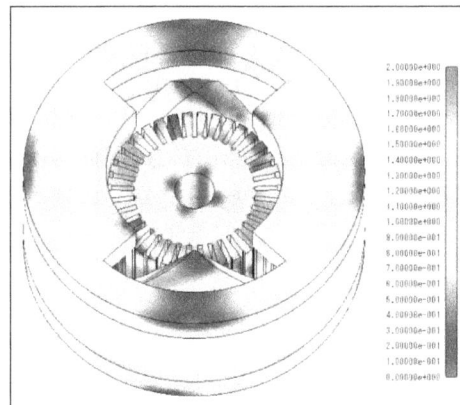

Contour of flux density.

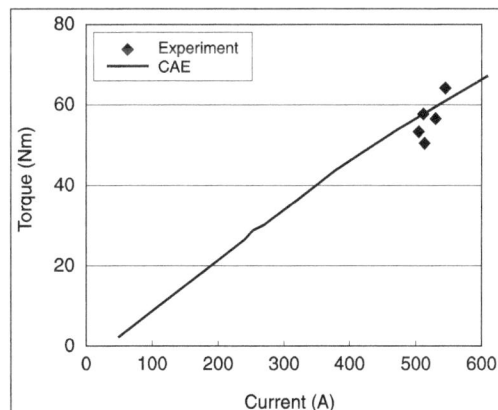

Torque curve of superconducting motor.

Construction and Appearance

Photograph below is an exterior view of the superconducting motor. The superconducting coil is immersed in liquid nitrogen inside the thermal insulation vessel, and it is necessary to compensate for liquid nitrogen evaporated during motor operation. To provide a continuous supply of liquid nitrogen during operation, liquid nitrogen reservoirs are included at the top of the motor.

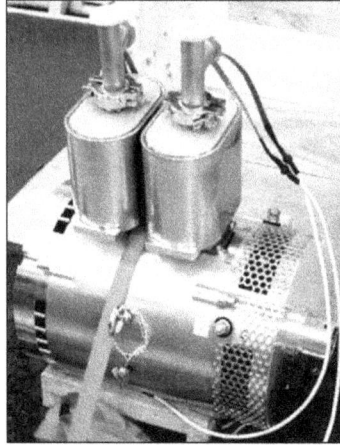

Superconducting motor.

Prototype Superconductive Electric Car

Successful operation of a golf cart using a bulk superconductor with an output of 2 kW. In the present study, the superconducting motor has been mounted on a commercial gasoline car (a Toyota Probox) so as to retrofit the vehicle as an electric car and verify its practicality in a vehicle environment that approximates normal use. The drive system of the prototype vehicle is diagrammed in figure. The power source for the motor is twelve 12-V lead batteries connected in series (144 V).

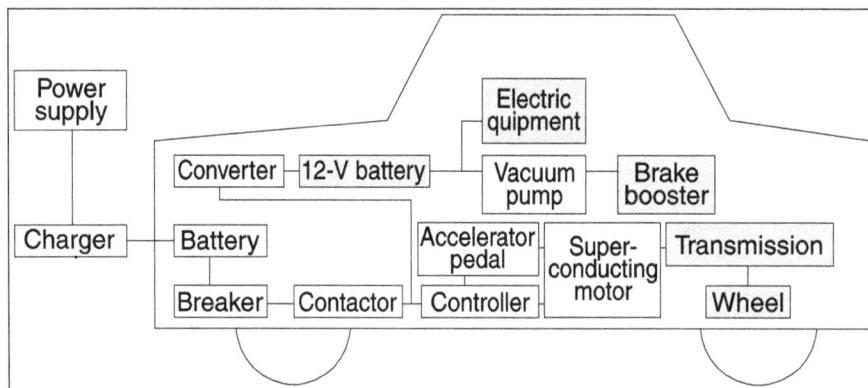

Configuration of superconducting car.

The pressure on the accelerator pedal is measured with a sensor, and a commercial current controller is used for supplying the corresponding amounts of current to the motor. A torque from the motor is conveyed to the wheels as drive power by means of the car's transmission. A contact-type charging method is applied, using a cord running from the external charger to a plug in the engine compartment. The removal of conventional engine also eliminates the negative pressure used by the brake booster, and therefore a commercial vacuum pump designed for electric vehicles is mounted on the prototype vehicle. A DC-DC converter is installed to supply power to 12-V electrical equipment, and the power-hungry air conditioner has been removed, leaving the vehicle quite similar in function to a basic gasoline car. The weight of the prototype vehicle is about 1,200 kg.

Photo below shows the engine compartment with the superconducting motor, current controller, and batteries installed. The motor's compact design enables transverse installation and it thus leaves ample room in the engine compartment for the current controller, the DCDC converter, the vacuum pump for brake assist, and two batteries.

The running evaluation results for the vehicle are shown in table. As is characteristic of electric vehicles, the highest torque is obtained at low speeds; hence, smooth acceleration from a standstill is possible even when the vehicle is in third gear. The maximum verified speed was 70 km/hr, and it is suitable for use in normal driving.

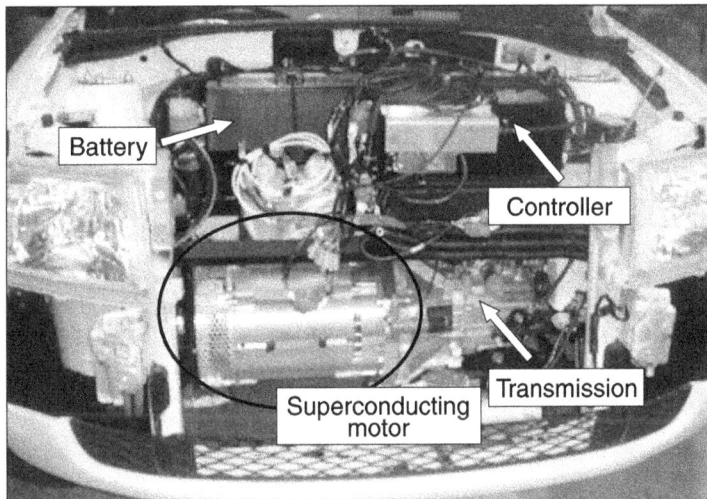

Superconducting motor fitted inside vehicle.

Table: Vehicle Performance.

Items		Performance
Vehicle Performance	Maximum torque (1,000 rpm)	70Nm
	Maximum output (4,500 rpm)	18kW
	Maximum speed (3rd gear)	Above 70 km/hr
	Cruising distance (at 30 km/hr on test site)	36km

Application in Transmission Lines

The linear SC excitation current coil - transmission line includes the several hollow cables developed for the Nuclotron magnets. The assemble view of the SC cable is given in figure. The 5*0.5 mm² copper-nickel tubes are wraped up by the 31 SC wires 0.5 mm in diameter having the 1045 SC

NbTi 10 μm filaments. The tube is covered by the insulation and nichrom wire band. The critical current of SC cable in the field of 2.5 T is 6.8 kA. The maximum operation voltage of insulation is 2.5 kV. The minimum bending radius is 18-20 mm. The cable has the high mechanical strength and stable to radiation up to 5×10^6 Gr.

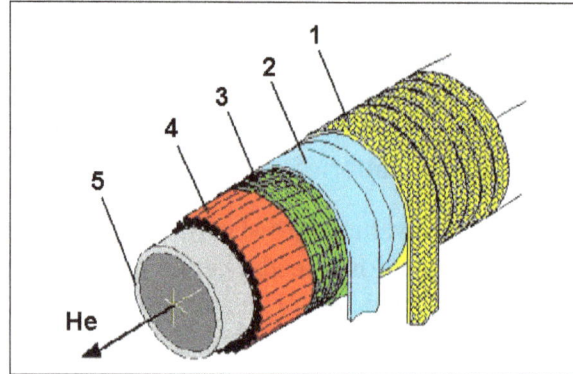

Nuclotron superconducting cable: 1 – glass-fiber tape, 2 – kapton tape, 3 – nichrom wire, 4 – superconducting wire, 5 – copper-nickel tube.

It is not efficient to use the very high excitation current in separate magnets of the transport beam lines because of the large number of SC current inlets. Beam transport system does not require the pulse operation mode. Then the DC SC magnets for that purpose were proposed. The electrically insulated SC wires are connected successively to form the winding with 200 A DC. The transmission line is designed following the standart of the long cryogenic systems technology "pipe in pipe". The screen vacuum insulation decreases the heat flow from the room temperature region to the SC cables. The magnet coil consists of 7 superconducting cables. The transmission line coil is placed inside a stainless pipe 21 mm in diameter and 0.5 mm in thickness. The screen-vacuum insulation of the coil pipe is made of 30 layers of 12 μm aluminized lavsan. The cable tube is mounted on the insulation support in the vacuum jacket. From the estimation, the heat leak to the superconducting lines is about 3 W/m.

Magnet Design

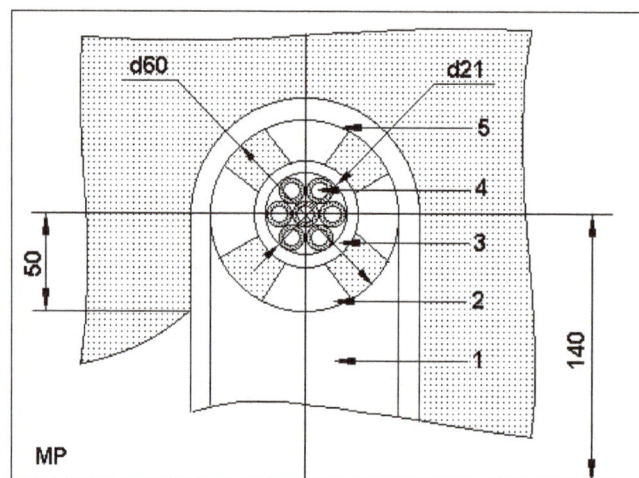

Superconducting coil for double transmission line magnet: 1 – coil support, 2 – insulating coil support, 3 – screen vacuum insulation, 4 – cables, 5 – vacuum jacket, MP – median plane.

The numerical calculation of the C-type transmission line magnets has been carried out. The optimizations of the yoke and pole profiles, field distribution, parameters of the SC line and mechanical strength have been considered for 1 and 2 line magnets (10 and 7 SC cables in each line).

The magnets provide the peak guiding field of 1.2 T and field gradient of 4.5 T/m. The forces in transmission line are $F_{x,y}$ = 3/0; 0.1/0.8 t/m correspondingly. The cross-sections of the magnets are shown in figures.

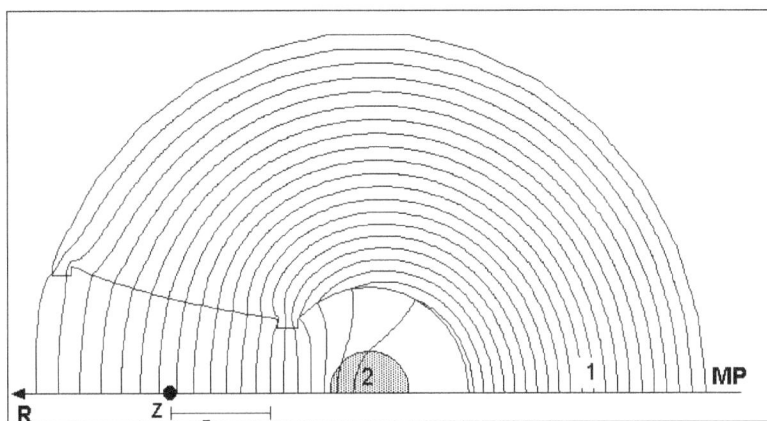

Defocusing transmission line magnet with 1 coil: 1 – iron yoke, 2 – superconducting coil, MP – median plane, Z – reference beam axis. Equipotential lines of the magnetic field are shown.

Booster Synchrotron

The magnetic field shaping is more effectivelly realized for 2 line magnet. This magnet was chosen for Nuclotron booster. The main parameters of F and D magnets are given in table. 63 m synchrotron includes 8 superperiods. Each of them consists of three doublets of combined function C-type magnets and one straight section. Drifts are used to install acceleration, injection, fast and slow extraction systems. The F and D magnet gaps are faced outside the ring for easier injection and extraction. The developed magnets allow to build the fast cycling, compact and economic synchrotron with the energy up to 500 MeV/u.

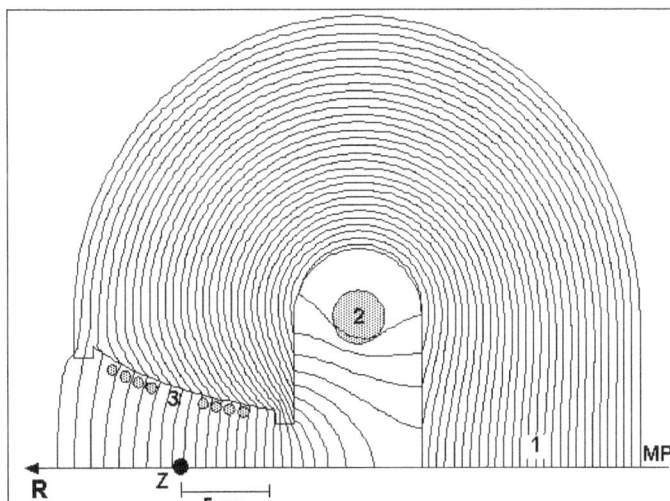

Defocusing magnet with 2 superconducting line coils and warm correction coils:
1 – iron yoke, 2 – superconducting coil, 3 – correction coils.

Table: Parameters of F and D magnets.

Length	1.0 m	N coils	2
Dipole field	0.9 T	N cables	7*2
Field gradient	4.5 T/m	Amp. turns	2*42 kA
Deflect. Angle	7.5 grad	Current	6 kA
Bending radius	7.6 m	Energy	8 kJ
Aperture	10*6 cm²	Inductance	0.5 mH
SC cable diam.	7 mm	Weight	700 kg

It could be used for multi-turn injection and storage in Nuclotron and also for scientific and applied investigations like radiotherapy with 450 MeV/u carbon beams. General synchrotron parameters are presented in table.

Table: General synchrotron parameters.

E:p inj/max	20/1200 MeV	Repetition	5 Hz
ions, A/Z=2	5/450 MeV/u	N super.	8
$B\rho$ inj/max	0.6/6.7 Tm	N FD/drf.	24/8
Perimeter	62.88 m	N mag.	48
L mag/drift	1/1.86 m	R bend.	7.6 m
B inj/max	0.085/0.88 T	Q'_x/Q'_y	-0.7/-2.4
G inj/max	0.4/4.2 T/m	$\beta_{x,y}$ max	9.2 m
$A_{x,y}$ $\pi\mu m$	400/250	D_x max	2.6 m
$\varepsilon_{x,y}$ inj/min	50/45; 5/4	Comp. fact.	0.2
Q_x/Q_y	2.4/2.45	$\Delta p/p$	±0.001

Beam Transport System

The wide distributed transport system of the Nuclotron extracted beams is based on the warm magnetic elements. Then the several experiments go in parallel the power consumption overcomes the Nuclotron ring consumption. There are two ways to solve the problem: removing the old transport system by SC elements - dipoles and quadrupoles with transmission line; removing only the coils of the existing elements by SC ones. Now the real way is to replace the old winding with 200-250 A DC SC coils. The new technology of SC cables should reduce the power considerebly.

Undulator System

The transmission line technology could be successfully used in FEL systems based on undulators - devices with the periodic magnetic field distribution: $B = B_0 * \cos(2\pi * z/\lambda_u)$, where, λ_u - period length of magnetic structure, B_0 - magnetic field amplitude. The energy of the electron bunch going in undulator is transformed into photon energy. The generated wavelength can be presented as $\lambda = \lambda_u * K^2/4\gamma^2$, where γ - relativistic factor, $K = 93.4 * B_0[T] * \lambda_u[m]$ - undulator factor.

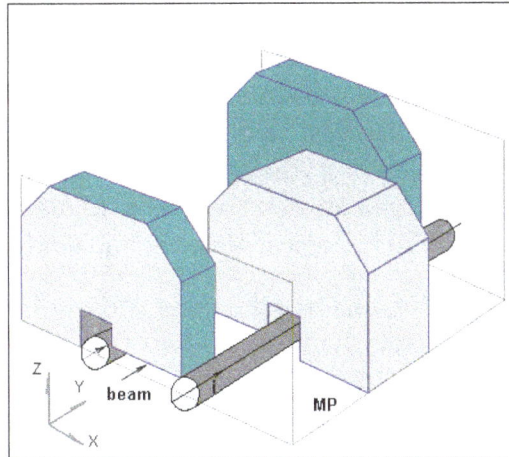

Scheme of the undulator period (upper part).

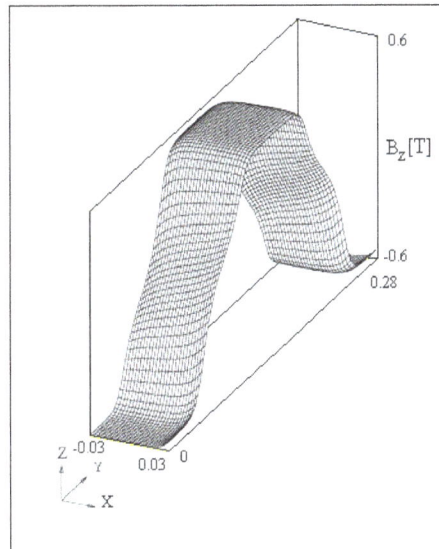

Magnetic field distribution in undulator median plane over beam excursion period.

For a wide range of FEL researches the wavelength variation is desirable. The possibilities to adjust wavelength are beam energy scanning, variation of the magnetic field and changing of the undulator periodicity. The permanent magnet undulator provides the short period λ_u and easier mechanical alingment. But long undulator systems require the big volume of the magnetic material and become very expensive. The electromagnetic undulator with SC transmission line is serious alternative to permanent magnet device. In figure the proposed design of 2 line undulator is shown. The total current of 6 kA in 7 cable SC transmission line produces the peak field of 0.6 T. With the beam energy of 800 MeV and undulator period of 0.28 m the radiation wavelength is 7 µm.

Application in Medicine

As new technological developments emerge, a logical move is to search for applications in medicine and healthcare. Superconductivity has been no exception. The primary application in this area has been high-field, very efficient magnets, with noteworthy examples in cyclotrons in radiotherapy and isotope production applications, beam delivery for reducing the size and operating cost for proton and ion beams for therapy, and most notably for magnetic resonance imaging (MRI) magnets. In the last case, currently the largest commercial application of superconductivity, these magnets have for all practical purposes been the principal enabler of this new imaging modality, which is having a tremendous impact not only on the quality of diagnostic information now available, but also in unraveling function in biological systems, and studying biochemical processes.

One particular medical application of superconductivity will not be addressed, as it is quite far afield from the main focus of this journal; namely, the use of Josephson junctions, primarily in the form of SQUIDs (Superconducting QUantum Interference Devices) as ultrasensitive pickups of electromagnetic signals.

Superconducting Cyclotrons as Sources of Medically Relevant Particle Beams

Few people have not heard of the Bragg peak as an argument why protons (or beams of other ions) offer substantial advantages for "external beam" radiation therapy. Figure schematically compares the dose distribution for such ions with similar beams of X-rays (or neutrons). As energy loss (related to energy deposition in the surrounding medium) varies as $1/E$, charged particles stiff enough to travel in a straight line (which electrons aren't!) will lose most of their energy as they stop at the end of their range. Multiple scattering does broaden the peak a bit, so heavier ions will show sharper Bragg peaks, but for the most part protons already can concentrate a stopping dose in a tumor volume, and provide very significant sparing of normal tissue compared to X-rays. This translates into fewer side effects, greater tolerance to the treatment and, particularly important for pediatric cases, a much lower chance of secondary tumor generation many years after the treatment.

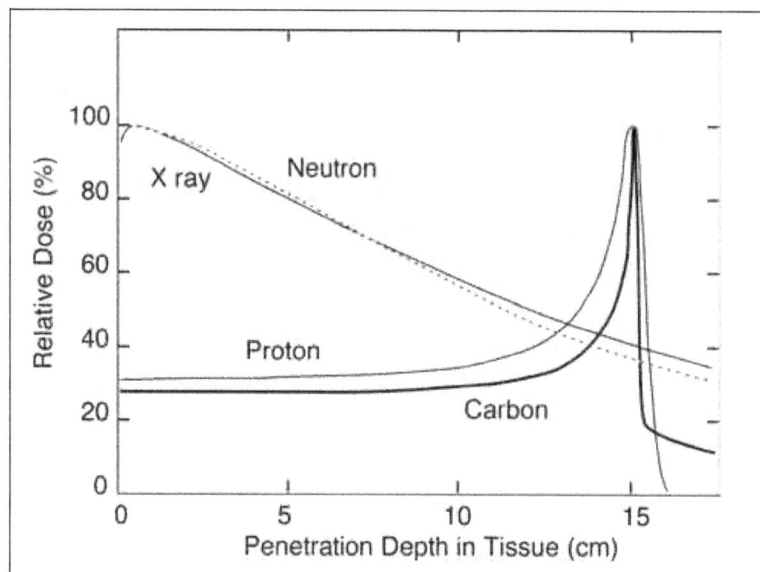

Figure shows comparative dose distributions for various therapy beams. Protons and carbon exhibit the "Bragg peak" dose enhancement in the vicinity where the particles stop, compared to the basically exponential dose distribution for X-rays. High-energy neutrons (~60 MeV) have similar energy deposition to the photons from a 7–8 MV source. The carbon Bragg peak is sharper owing to less multiple scattering. The dose for carbon beyond the end of the range is from lighter (longer-range) fragments from nuclear reactions of the slowing carbon ions.

We are currently seeing a tremendous growth in the number of centers offering proton therapy, and now overseas (but not yet in the US) a growing number of hospitals and research institutions delivering carbon ions. Not only is the Bragg peak sharper for carbon, allowing finer control of field edges and distal falloff, but the higher charge of the carbon ion induces greater biological damage at the stopping point. This factor, termed LET, varies roughly as the square of the charge of the ion, so each carbon ion (Z = 6) deposits about 36 times the energy at its stopping point as a proton.

Clinical practice for the treatment of cancer has developed around the ability of tumor tissue to repair from radiation "insults," and the large number of fractions usually given in a course of radiation treatments (as many as 48 daily fractions) arises from the slightly different repair rate between normal and cancer tissue (cancer normally repairs more slowly).

Carbon ions change this paradigm. By introducing more ionizing radiation into the cells, more damage is done and much less repair is possible.

Neutrons have been used for radiotherapy for almost 50 years as well and have shown effectiveness in certain sites. A fast-neutron radiation field is usually produced by sending 60 MeV protons or deuterons onto a beryllium target, and the resulting neutron depth–dose curve is approximately equivalent to that of a 7–8MeV photon beam. The dose is delivered by the neutrons transferring energy to protons though nuclear collisions; the recoiling (low-energy) protons break molecular bonds and do the biological damage. In fact, the LET of these recoiling protons is substantially higher than X-rays, so there was a high degree of optimism in the early days that neutrons could provide substantial improvements over X-rays for cancer treatment. Figure shows neutrons as having higher biological effectiveness, but from figure we see that the dose distribution is similar to X-rays.

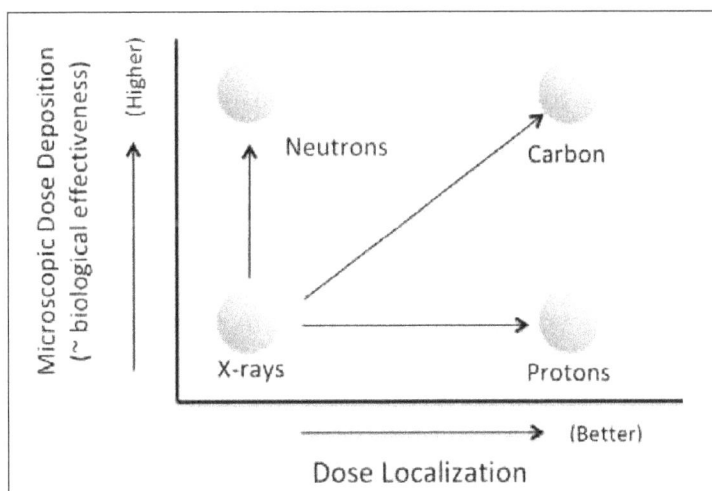

Comparison of X-rays, protons, neutrons and carbon ions for relative effectiveness: X-rays have low LET and poor dose distribution, and protons also low LET but good dose definition; neutrons have poor dose definition but higher LET, and carbon offers both high definition and high LET.

In fact, the high expectations for neutrons did not materialize; the clinical results were dominated by complications owing to the lack of good dose localization, and excessive damage to surrounding normal tissue.

So why this long introduction to neutron therapy? Simply that the first use of superconductivity for clinical therapy was a neutron-producing cyclotron designed and built by Henry Blosser and installed at the Harper-Grace Hospital in Detroit.

Clinical Requirements for Charged Particle Beams

Table below details the three clinical parameters most relevant for the provider of ion beam acceleration and beam delivery.

Table: Selected clinical requirements for ion beams for external beam therapy.

Beam range in tissue	30 cm
Beam current (average)	Adequate to treat 1 L volume to 2 Gy in 1 min
Delivery orientation	Isocentric around patient

Range is determined by need to reach any anatomical site, and as the target is usually treated with overlapping fields, beam must enter, and reach the target volume, from several orientations. To guarantee 30 cm in the patient, the usually specified energy for protons is 250 MeV. While this corresponds to a range of almost 37 cm, energy is lost in the devices used for forming the large field, particularly if passive scattering is used. Scanning systems have less material in the beam, and a lower top energy could be provided from the accelerator and still meet the 30-cm-range requirement in the patient. The accepted beam energy for carbon ions is 400 MeV/amu.

Beam current is specified in terms of treatment time. Delivering a fraction in a minute or less is desirable from a patient comfort standpoint — immobilization devices are rarely conducive to a patient remaining still, as well as for promoting efficient patient throughput for the treatment room. It is difficult to specify an actual beam current, as the particle flux required for the prescribed dose depends on the size, shape and depth of the field, even for targets of equal volume. For a proton beam interfaced with a scanning system, an average current of a few nanoamperes is usually adequate. However, a cyclotron is a fixed energy machine; beam is always extracted at its top energy. To place stopping particles at different depths in the target, the energy needs to be varied. With beam extracted from a cyclotron this is done with a wedge-degrader and energy selection magnet system. In addition, a set of jaws selects the beam of correct emittance to pass through the gantry. Because of multiple scattering, greater energy reduction leads to large beam losses on the jaws, and so while at full energy most all of the beam is acceptable for downstream transport, at 100 MeV this efficiency could be a few percent or less. Thus, to meet the treatment time specification the beam current from the cyclotron must be increased, to preserve the "brightness" of the beam as it is transported to the gantry. To complicate matters, substantial beam loss can occur in the process of converting the very tight beam entering the treatment room to the radiation field covering the size (the maximum field can be 20 × 20 cm or greater) of the target. Passive scattering, for instance, involves at best only 40% transport efficiency, again increasing the need for more particles from the accelerator to meet the treatment time requirement. In all, cyclotron currents for a viable therapy application must approach the microampere level.

Treatments are normally given with patients lying on their backs on a treatment couch. This position usually provides the greatest comfort for the patient, but more importantly corresponds to the usual position in which diagnostic information is obtained. With high precision beams, having very accurate information of the actual coordinates of the target volume is essential. Unfortunately, the human body is far from being a rigid structure and the best chance of knowing where a target is located will occur if treatment is performed in the same position as the diagnostic scan was made.

The implication for beam delivery is that beam should be able to enter the patient from any angle over a full 4π sphere. The most logical way of achieving this is to bring the (horizontal) beam into the treatment room, and run it through a "gantry," a rotating transport line that can put the beam at any angle in a vertical plane with the target volume located at the "isocenter." Oblique entry ports are obtained by rotating the patient couch in the horizontal plane.

X-rays for therapy are produced now by compact electron linacs that can easily be accommodated on such a gantry, or in any event the electron beams of 20MeV or less can be bent in ~ 10 cm radii by reasonable magnets. Achieving "isocentric delivery" with electron beams is quite practical.

However, the magnetic rigidity ($B\rho$) of a 250MeV proton beam is about 2.5 tesla-meters (T-m), so gantry sizes with conventional magnets become very large. 400MeV/amu carbon beams have a $B\rho$ of 6.6 T-m, making isocentric delivery even more difficult.

Role for Superconductivity

The first machines used for investigation of protons and ions for therapy were accelerators in nuclear physics laboratories where, though not under ideal conditions, the groundwork was laid for transferring proton- and ion-beam therapy to the hospital environment. The first hospital-based proton facility was installed in 1990 at Loma Linda California, and consisted of a Fermilab-built 250MeV synchrotron and three large gantries.

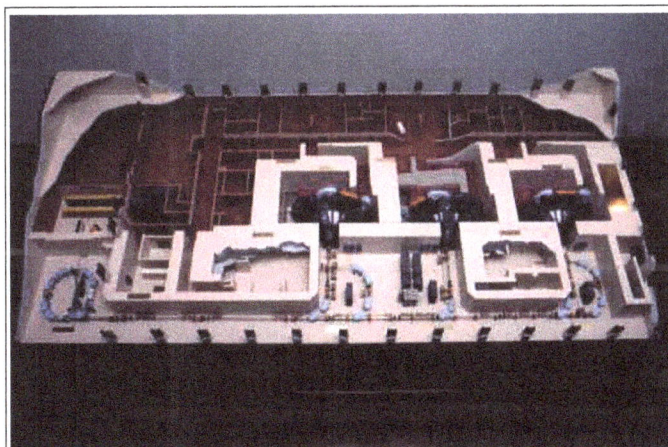

Model of the Loma Linda proton therapy facility. The overall facility dimensions are approximately 90×40 m. A compact, ~ 7-m-diameter weak-focusing synchrotron (lower left corner) provides 250 MeV proton beams to three gantry rooms and two fixed beam rooms. Over 15,000 patients have been treated since 1990.

The diameter of each gantry is 13 m. In 1993 IBA won the contract for providing a 235 MeV isochronous cyclotron with two gantries plus two fixed beam lines to the Massachusetts General

Hospital. This configuration — an accelerator (normal-conducting) and a backbone beam transfer line to several treatment rooms, most of them provided with gantries — has become the standard for essentially all of the new treatment centers. In the US alone, IBA and other vendors have installed eight facilities modeled after the first MGH unit (the number is growing almost monthly), and growth in other parts of the world has been equally prolific.

The rationale for this configuration is that the accelerator is the most expensive component, and the use of beam in each treatment room is very low (most of the time it is used in patient setup and alignment), so a single accelerator can service several — as many as five — treatment rooms. The corollary of this, though, is that a center for delivering proton (or other ion) treatments will be very large and very expensive. Costs for new proton facilities of this ilk are now on the order of US$150 million.

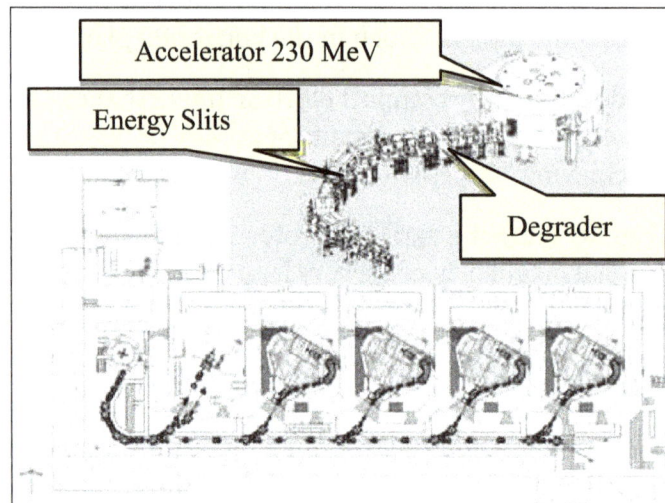

IBA's four-gantry layout. The cyclotron vault is at the far left; the graphite degrader and energy selection spectrometer (highlighted in the inset) passes beam into the long transport line, where it is peeled off into the appropriate treatment room. Each gantry weighs about 100 tons, and is 12 m in diameter.

While an argument can be made that this is not an unreasonable price for building a new radiation therapy center capable of treating over 1000 patients per year, it is nonetheless a substantial hurdle for the vast majority of hospitals that might be interested in providing proton therapy and that cannot justify a facility of this size.

The drive, then, is to reduce the cost of hardware, delivery systems, and facilities to make protons more affordable. If inexpensive sources of 250 MeV protons could be built, and gantry size reduced to not require a huge vault, then one- or two-room systems could be within the reach of many more customers. The ultimate goal would be to reduce the size to where an existing vault currently housing an electron linac could be upgraded to contain a proton system.

The most effective way of achieving a more compact size is to increase the magnetic field in the accelerator and delivery systems, using superconducting magnets.

Cyclotrons for Therapy

Isochronous cyclotrons work correctly when the magnetic field design is valid for a given ion and final energy. Since their introduction in the late 1950s, however, resistive-magnet-based cyclotrons designs have been challenged by the nonlinearity of the pole and return yoke steel. In a typical project, a succession of model magnets would be built to validate and optimize the magnetic field configuration before the design of the actual machine could be finished. It was realized in the early 1970s that, by substitution of the standard hollow copper conductor-based resistive coils in isochronous cyclotrons with superconducting coils having higher current density and substantially more ampere turns for a given size, the iron in the poles would saturate and become computationally linear. This meant that for the first time one could design computationally an isochronous cyclotron, and build it with the expectation that it would work as designed. Quickly, it was shown that one had to make a choice between high bending strength and high flutter — the former yielding heavy ion machines and the latter leading to energetic proton machines. The engineering superconductor of choice at that time being NbTi led immediately to cyclotrons in the 3–5 T range. The first of these was the variable energy K500 heavy ion cyclotron at MSU. At 5 T, the K500 is roughly one-tenth the overall size and mass of an equivalent resistive-magnet-based cyclotron, and consumes roughly one-third the wall plug power, including the cryogenic systems.

In addition to making the machines more compact, there were operational advantages to going superconducting: reduced thermal cycling, less magnet charging hysteresis, smaller/lower power RF systems, and wider tuning ranges in the variable energy machines. However, this came with new engineering challenges: complex cryostat designs, higher magnetic forces, handling cryogens and quench protection systems. But, once commissioned, these cyclotrons have lifetimes measured in decades; all have met their performance goals, and cryogen plants have been replaced by mechanical cryocoolers in the latest systems. The K500 produced its first extracted beam in 1982, and is still in routine operation at the present time at MSU.

The emphasis here on the MSU K500 is not an accident. All of the superconducting cyclotron designs now being developed for cancer therapy share most of its overall features. An attractive configuration at the present time has the cyclotron closely coupled to the gantry, providing beam to a single or just a pair of treatment rooms. This significantly reduces the cost of a therapy facility and is seen as a key to wider deployment of proton therapy systems. The Mevion Monarch (superconducting synchrocyclotron), the Varian Medical Systems PROSCAN (superconducting isochronous), both in operation, and the upcoming IBA S2C2 (superconducting synchrocyclotron) are all based on this configuration. The Varian ProScan, first built by ACCEL, now a division of Varian Medical Systems in Germany, has also been installed in the until-now-standard multiroom configuration. But, before describing these systems, the first superconducting machine to deliver therapeutic beams.

Neutrons: MSU/Harper-Grace

Henry Blosser and Dr. William Powers with the completed
neutron source cyclotron mounted on a rotating gantry.

In the 1980s, it was not as clear as now as to whether particle therapy should be done with protons, neutrons or heavy ions. Some medical groups in fact favored neutron therapy. In 1984, Henry Blosser started the design of a superconducting 50 MeV deuteron cyclotron, under a commission from Dr. R. L. Maughan, as a neutron therapy instrument for the Wayne State University Harper-Grace Hospital in Detroit, Michigan.

First operated in 1992 with patients, it ran routinely until about 2007, when it was shut down for programmatic reasons. An attempt was made to recommission it in 2010, but this effort was scuttled by a hard-to-diagnose vacuum leak that had developed in the cryostat. To date, this machine remains the only superconducting cyclotron to be developed for neutron cancer therapy.

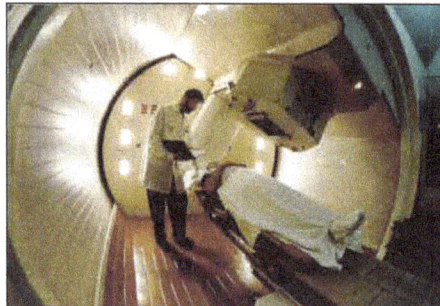

Patient's eye view of the neutron source. All equipment is hidden behind the cowling. Cyclotron rotation covers 360°; movable floor planks rotate out of the way when the cyclotron is below floor level.

The upper yoke and poles for the 50 MeV deuteron cyclotron, without the cryostat. The neutron exit channel is at the top. The maximum field is 5.5 T, the average field is 4.6 T, and the beam radius when it strikes the target is 45 cm.

This compact superconducting isochronous cyclotron established many key features now seen in new machines. It was gantry-mounted, providing a collimated secondary neutron beam to a single treatment room. Since this machine was cooled with both liquid nitrogen and helium, a novel cryostat design was developed — and is now patented — which would insure that the cryostat boiloff vent lines were always above the level of the cryogens in the rotating magnet. It was a Q/A = 0.5 cyclotron, accelerating deuterons to 50 MeV total energy (25 MeV/amu), which impinged on an internal beryllium target (thus sidestepping the need for an extraction system) to generate a secondary neutron beam. Magnet steel helped harden the exiting neutron spectrum, and tungsten collimators in the yoke concentrated the neutron beam on the isocenter. The high radiation generated inside the cyclotron required the selective use of tungsten shielding in the cryostat to protect the superconducting coils. The field design was similar to that of the K500, with a 5.5 T maximum field on the hills and a 4.6 T average accelerating field. With a compact high field central region, it employed a 10 μA internal Penning ion source with separate cathode feeds on opposite sides of the median plane. This ion source concept is still in use in the Varian PROSCAN cyclotron.

Protons

The focus of development in recent years has been on compact cyclotrons for delivery of proton beams. In his 1988 report describing his deuteron machine, Blosser already envisioned a super-conducting machine mounted on a gantry counterbalancing the transport magnets bringing the beam to the patient. Systems from Varian and IBA are now realizing almost exactly this same picture. Both isochronous cyclotrons and synchrocyclotrons have been developed for this application.

Varian Medical Systems/ACCEL PROSCAN Isochronous Cyclotron

Long-term studies of the efficacy of proton therapy by the Radiation Oncology Department at the Massachusetts General Hospital (MGH), Boston, conducted at the Harvard Cyclotron Laboratory's (HCL) old 165MeV synchrocyclotron, provided strong motivation to acquire a dedicated proton therapy facility at MGH. The National Cancer Institute (NCI) funded a study grant, jointly conducted by MGH, HCL and the Lawrence Berkeley National Laboratory (M. Goitein and J. R. Alonso, PIs), which led to development of very complete clinical specifications. NCI, as a condition for providing initial construction funding for a facility at MGH, stipulated that the machine and technical components for the system would be procured from the private sector, not from a national laboratory.

In response to a call issued by MGH, proposals were received for both cyclotron- and synchrotron-based systems. One of the proposals included a superconducting isochronous cyclotron design developed by Blosser and his MSU team. While MGH selected the normal-conducting C235 isochronous cyclotron from IBA, work continued on Blosser's concept.

The overall magnet design followed the K500 example, except that the peak field was limited to 3T in order to have sufficient vertical focusing. At 3T, this machine is about half the mass of the IBA C235 proton therapy cyclotron, so it is taking advantage more of the other features of super-conducting cyclotrons: linear iron, minimal thermal cycling and low electrical wall power. ACCEL GmbH began to work with MSU on the machine in 2001, with construction of the first system starting in 2003 as the centerpiece of the PROSCAN Project at the Paul Scherrer Institute (PSI) in Villigen, Switzerland, aimed at providing a dedicated source of protons for the PSI medical programs. Prior to this, protons were obtained by slicing off a tiny fraction of the 2mA beam from their megawatt-class 590MeV ring cyclotron, and degrading them to 200MeV at the entrance of the medical beam area. In conjunction with PSI, and with support from MSU, ACCEL began installation of the first production system in Villigen at the end of 2004. A key innovation was the elimination of the cryogen plant, in favor of a closed loop liquid helium refrigeration system operating in the thermal siphon mode with a set of four dedicated cryocoolers. An additional single-stage cryocooler provided cooling for the intermediate temperature thermal shield in the cryostat. The PROSCAN facility uses an energy degrader and energy selection system in the transport line to the treatment area to provide variable energy from 70 to 250 MeV. Intensity variations, to control the dose rate, are accomplished by means of a vertical deflector plate at the center of the cyclotron. This accurate intensity modulation coupled with a gantry with transverse beam steering allows precision spot scanning, resulting in a precise 3D dose distribution in the tumor while sparing nearby normal tissue. The pioneering scanning development work done at PSI has stimulated most system manufacturers to develop their own scanning systems for proton therapy.

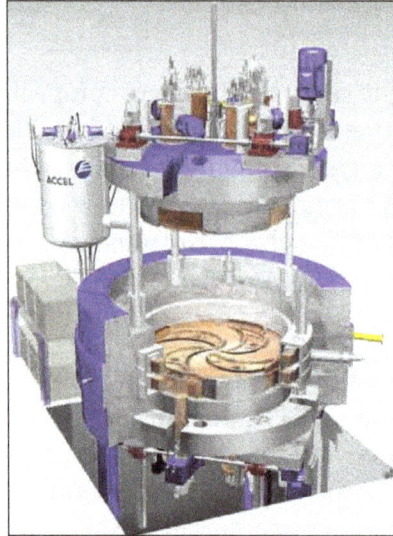

CAD cutaway of the ACCEL PROSCAN isochronous 250MeV proton cyclotron with the pole elevated.

Enlarged view of superconducting coils for the PROSCAN cyclotron. NbTi coils.
The current is 160 A, corresponding to about 106 ampere turns.

The switch of the clinical programs to the PROSCAN system occurred in late 2006, with the first patients treated with proton beams from the superconducting cyclotron in February 2007.

While the first PROSCAN system was being fabricated for PSI, ACCEL contracted to build a second identical machine for Munich's Reineker Center. The simultaneous assembly of two systems led to numerous challenges, with the Munich system lagging behind the Villigen system significantly in final acceptance. The first patient treatments occurred in Munich in March 2009.

Varian Medical Systems acquired all of ACCEL in 2007, and sold the research instruments component of ACCEL to Brucker in 2008, in order to concentrate on the development of proton therapy systems.

In 2012, a new dedicated Varian proton beam therapy fabrication center is fully operational in Bergish Gladbach, and five additional cyclotrons are in various stages of assembly there, with unit No. 3 recently installed at the six-treatment-room proton therapy facility under construction at the Scripps Medical Center in San Diego.

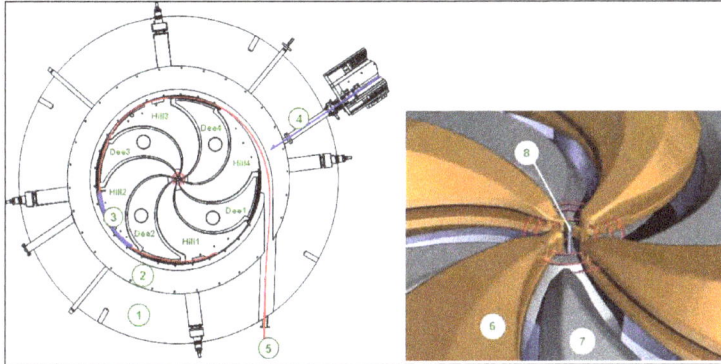

Midplane cut and central region of the ACCEL cyclotron: (1) iron yoke ring, (2) cryostat containing a superconducting coil, (3) extraction electrode, (4) retracted radial beam probe, (5) extracted beam, (6) accelerating dee, (7) field-shaping magnet pole ("hill"), (8) ion source. In red are the first spiral orbits in the central region and the beam extraction path.

Mevion/Still River Synchrocyclotron

The Harvard Cyclotron was in fact a synchrocyclotron, which is weak-focusing like the original Lawrence cyclotrons, but addresses the relativistic mass increase of the proton by letting the resonance frequency fall synchronously during acceleration. The main guide magnetic field is obtained from an azimuthally symmetric pole that creates the required weak focusing to stabilize ion motion during acceleration. In contrast to isochronous cyclotrons, which rely on strong focusing, derived from the "flutter" created by an azimuthally varying magnetic field, no flutter is required. This is important since, to zero order, flutter scales inversely with the average field. Note that the long acceleration cycle, on the order of microseconds, means that intrinsically synchrocyclotrons are low-duty-factor and hence lowintensity accelerators. However, the experience at Harvard showed that the proton current that could be extracted from a synchrocyclotron was more than adequate for proton therapy.

Using the magnet technology developed for the superconducting isochronous cyclotrons by the mid-1980s, one could get to a 3T superconducting isochronous cyclotron, with the limit principally being creating sufficient flutter for stable vertical motion. This leads to cyclotrons of between 50 and 90 tons at 250MeV. To make them smaller and more compact, one has to operate at even higher magnetic fields. To overcome this limitation Blosser and his team at MSU looked at the possibility of reintroducing the synchrocyclotron, first in a set of studies leading to a patent for a 5T synchrocyclotron in 1985, and later in the thesis study by Xaio Yu Wu, which addressed in more detail fundamental acceleration issues which demonstrated that such a superconducting synchrocyclotron was indeed feasible.

Little changed with respect to the feasibility of compact high-field superconducting synchrocyclotrons until Ken Gall and Timothy Antaya met in January 2003. Gall, which had earlier been radiotherapy postdoc at Harvard, was looking for a way to realize a single-treatment-room concept, like the neutron therapy installation at Harper-Grace Hospital discussed above, but for protons at 250 MeV. To make a cyclotron on a gantry compact enough to fit in an affordable treatment room, Gall thought that he needed a synchrocyclotron above the 5.5T limit set by the previous MSU studies. Antaya, who had just arrived at MIT, had been working on large-scale superconducting magnet systems for fusion and other applications, and had developed a broad array of new engineering techniques for the use of advanced superconductors in difficult high-field magnets. Antaya thought that he knew how to approach the problem of scaling both the beam dynamics and the magnetic field design of synchrocyclotrons to very high fields.

Gall and Antaya began a collaboration in 2003. By July 2004, Antaya and his team at MIT had demonstrated the feasibility of synchrocyclotrons at 8T and 9T, while Gall had formed Still River Systems to take on the challenge of the design and commercialization of a single-treatment-room proton therapy system based on the MIT synchrocyclotron design. The higher-field synchrocyclotron, at about 9T, was chosen as the baseline. It required the use of an advanced superconductor, Nb3Sn, which is brittle in the state where it is superconducting, but was a specialty of the Magnet Group of Joe Minervini at MIT. The effort to develop the 9T synchrocyclotron moved to detailed engineering in 2005–2006. Still River Systems began fabrication of the first synchrocyclotron in 2007, with the first extracted beam achieved at clinically required levels in May 2010. By late 2011, the first system had been installed at the Barnes-Jewish Hospital in St. Louis, which was FDA cleared for patient treatment in Spring 2012. Still River Systems became Mevion in 2011. At present, multiple systems are under fabrication at Mevion facilities, and it is possible that current orders will quickly result in a doubling of the number of superconducting cyclotrons operating worldwide.

The 9TMIT synchrocyclotron, shown in figure, includes a number of design innovations required to achieve the high-field particle accelerator, making it the most compact 250MeV proton accelerator built to date by any accelerator technique. The overall properties of this cyclotron are: a magnet diameter of 2m, a magnet height of 1.5m, a mass of less than 20 tons; the final energy is approximately 254MeV and the RF spans 100–150MHz. Ions are accelerated to full energy in about 16,000 turns taking 200ms, and a repetition rate of up to 1000MHz was planned. The central magnetic field is of order 8.9T and the peak field in the conductor is about 11 T, making this the highest-field circular particle accelerator for protons. The coils can be dry — no cryogens — cooled by a set of mechanical Gifford–McMahan cryocoolers. Ions reaching the full radius have a small momentum spread, allowing self-extraction via a passive fixed magnetic field perturbation, with full energy being achieved at greater than 90% of the maximum pole radius of around 30cm.

Ken Gall and the Monarch 250 8.9T synchrocyclotron.

Figure shows the single-room Mevion gantry configuration. The synchrocyclotron, known as the Monarch 250, is mounted so that the beam plane is vertical at the center of the "barrel" structure between the end arms, and the extracted beam is aimed directly at the isocenter. The energy-selection, collimation and dosimetry devices are mounted on a separate, substantially less massive rotating structure, rotating in synchronization with the main gantry. This lighter inner gantry allows greater beam positioning accuracy by avoiding the slight deflection errors introduced by rotating the outer gantry.

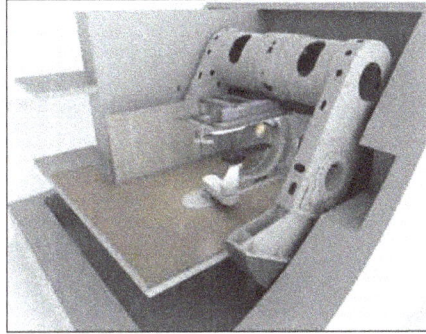

Mevion gantry system. The cyclotron is mounted directly in line with the patient couch, and the extracted beam points to the isocenter. The gantry rotates through 190°.

IBA S2C2 Synchrocyclotron

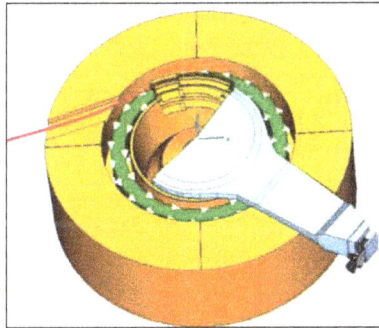

2009 concept for IBA's 250MeV synchrocyclotron.

In 1996, IBA began looking at superconducting synchrocyclotrons for proton radiotherapy independently of Still River Systems and MIT. In this case a fixed machine was desired, so high compactness was not required. In addition, the term of the Blosser–Milton superconducting synchrocyclotron patent, which set a field threshold for new noninterfering machine design of higher than 6.5T, had expired. With these considerations in mind, IBA settled on a synchrocyclotron concept with a field of about 6T, which would be significantly more compact, lower-cost and lower power than any proton therapy cyclotron existing at the time, including its own C235, developed for MGH, which by then was installed in a number of proton centers around the world. IBA used its own development team for the overall system design, magnet and the RF system for the S2C2 shown in figures, and contracted Pierre Mandrillon's AIMA Company in Nice, France to complete the design of the synchrocyclotron acceleration mode and ion source. Design work was completed in 2010. To set the scale of this compact superconducting synchrocyclotron, the completed magnet is shown in figure while other figure shows for size comparison the pole for the C235, IBA's normal-conducting isochronous cyclotron. As of early 2012, the magnet for the first system has been successfully operated at full field, while other components are well advanced, and the cyclotron is expected to produce beam by early 2013. The full clinical system, including pencil beam scanning and a compact gantry (but with normal-conducting magnets), is called Proteus One, and is shown schematically in figure. The first installation of this system will be at Centre Antoine Lacassagne in Nice, France, with patient treatments expected to start in 2015.

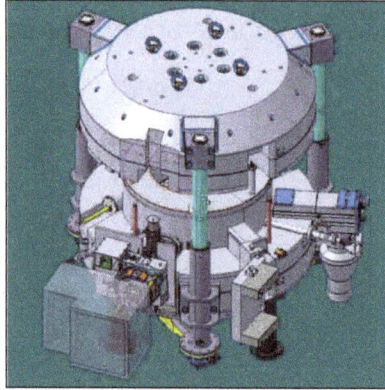

Schematic of IBA's S2C2 synchrocyclotron.

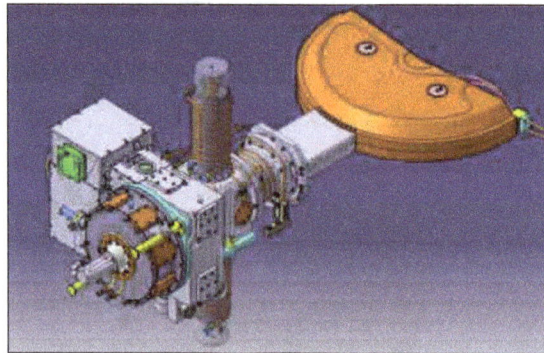

RF system for the S2C2 synchrocyclotron, showing the single dee. Frequency variation is accomplished with a rotating capacitor in the square box off the left side.

Ions

Yves Jongen with the magnet of the S2C2 250MeV IBA synchrocyclotron. Inset: ~1-m-diameter pole piece.

Pioneering Bragg peak therapy with ions heavier than protons was done in the 1970s, and '80s at the Lawrence Berkeley National Laboratory, first with the 184" synchrocyclotron and 225MeV/amu helium ions, then with a variety of ions from carbon to argon at the Bevalac. This accelerator had capabilities for achieving a 30 cm range in tissue for any of these ions, but the heavier ions experienced large losses due to nuclear reactions, so the beam — contaminated with nuclear fragments — was anything but pure when it reached the distal end of long-range fields. However, for shallower fields the Bragg peak was quite clean, and for ions up to neon, the fragmentation tail seen in figure at full range was acceptably small. The rationale for the very heavy ions, proposed by Tobias, was that biology experiments showed the so-called oxygen enhancement ratio (OER)[e] to be unity for silicon and argon. The medical team opted to do the majority of the Bevalac treatments with neon ions. Although the clinical results for the 433 patients treated between 1976 and 1992 were quite good, late effects seen in tissue along the entrance channel (the so-called "plateau" region upstream of the Bragg peak) suggested that neon ions may have been too heavy. This has led to carbon being the ion of choice for all follow-up investigations, now occurring in Japan and at several centers in Europe. A total of almost 8000 patients have now been treated with carbon ions at these centers.

Steel yoke for IBA's C235 normal-conducting 230MeV proton isochronous cyclotron. The diameter is almost twice that of the S2C2, and the weight four times as much.

To date, accelerators for carbon ions have all been quite large synchrotrons, following the example of the Berkeley Bevalac and the GSI SIS-18 in Darmstadt. The operating facilities in Japan (HIMAC, Hyogo and Gunma) and in Europe (Heidelberg, Pavia) all have synchrotrons of ~20m or larger diameter, and transport lines bringing beam to three or more treatment rooms.

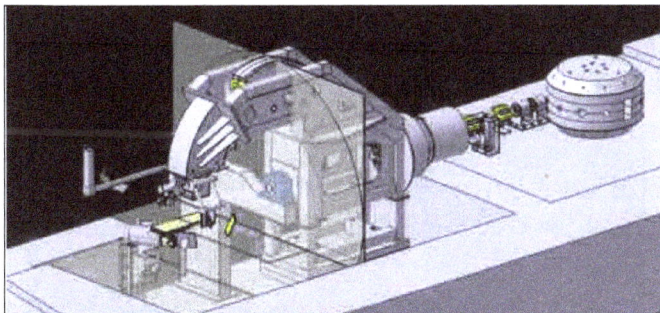

Schematic drawing of IBA's single-room Proteus One system.

Clinical research is driving the requirement that treatments with carbon be closely cross-compared with protons, so the new facilities all have the requirement of an accelerator capable of delivering both 400MeV/amu carbon (fully stripped 6+ ions) and 250MeV protons. The very large difference in rigidity of these beams creates an interesting challenge to the accelerator designer.

In all, the size and cost of these facilities are substantially greater (at least a factor of 2) than for a comparable proton-only facility, making the financial barrier even more daunting for those centers interested in exploring carbon as a therapy beam.

Decreasing the size and cost of the technical components is clearly a highly desirable goal. While the beam delivery challenges of ions, in particular the isocentric requirement in table1, are substantial, strong incentives exist for developing more compact sources of ion beams of suitable quality.

The first design effort for a superconducting neon cyclotron, namely the 1985 EULIMA project spearheaded by Mandrillon, was not successful in obtaining funding support owing to its size and complexity. It was also substantially before its time, coming when the radiation oncology community was only learning about the possibilities of using ions for therapy, and cost–benefit analyses could not be favorable, because benefits were still uncertain.

IBA C400: Isochronous Carbon/Proton Cyclotron

In the past ten years, studies on using cyclotrons to produce carbon and proton beams suitable for therapy have been conducted at Catania and IBA in partnership with JINR Dubna, both studies concluding that such cyclotrons were indeed practical, and that the two beam species could be produced in the same machine. The IBA C400 has reached a stage of maturity where funds are being sought to build the first unit.

The single-stage IBA C400 isochronous cyclotron utilizes axial injection of fully stripped carbon (up to 3 emA of C^{6+}), from ECR sources marketed by Pantechnik in Bayeux France, and of molecular hydrogen in H_2^+ form, from a multicusp source. The axial injection line can accommodate more than these two sources, so any ion species that can be produced with a Q/A of 0.5 can be given to the cyclotron for acceleration. As all injected ion species have the same charge-to-mass ratio, the basic acceleration parameters are similar, and only a slight change in the 70MHz (fourth harmonic of cyclotron frequency) is needed to accelerate any of these species.

The cyclotron, shown schematically in figure, has a 6.4m outer diameter, is 3.4m high and weighs about 650 tons. While quite large, it is still substantially smaller than the existing synchrotrons used for producing therapy beams of carbon. The pole radius is 1.87 m; the hill field is 4.5T, the valley field is 2.45T and fields at the coil are well within NbTi capabilities. There are plans to cool the coils with two cryocoolers running a closed loop helium thermosyphon system. Figure 20 shows the four highly curved hill segments, with RF cavities filling two of the valleys.

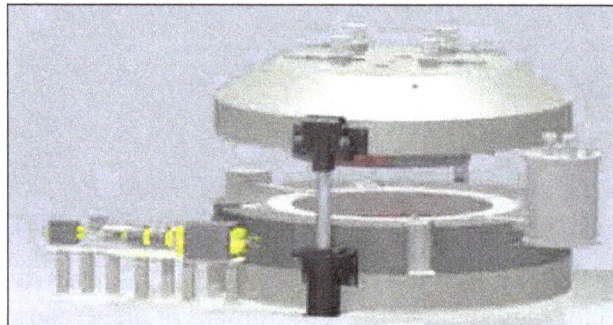

IBA's C400 isochronous superconducting cyclotron for 400MeV/amu carbon ions and 260MeV protons. The outer diameter of the steel is 6.3m.

The fully stripped species (e.g. C^{6+}) reach a maximum energy of 400MeV/amu, and are extracted via an electrostatic deflector; extraction efficiency of 80% or better is anticipated. The H_2^+, on the other hand, is given to a stripper foil placed where the beam reaches 260MeV/amu. The breakup of the H_2^+ molecule produces two protons that spiral inward (instead of outward, as is the case with H− cyclotrons), but the azimuthal field variation can be used to produce orbits that will allow the protons to escape. Figure shows that at the indicated foil placement the protons will undergo two spirals and exit at the same place as the carbon beam does. The angle at the extraction point is not exactly the same, so a chicane is employed to bring the two beams into a common transport line.

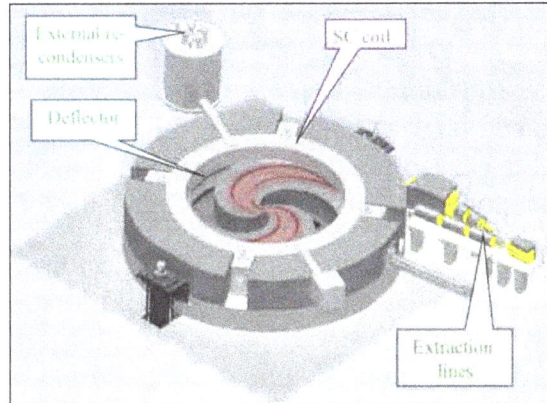

Cutaway of the C400 showing four spiraled hills and two RF cavities. Two extraction lines are seen; the outer one captures protons while the inner one transports Q/A = 0.5 ions. The last dipole brings the two beams into a common transport line.

The plan is to use the same energy selection and transport system which IBA employs with its proton systems, as shown in figure. Calculations indicate that degrading carbon beams is not unacceptable; the nuclear breakup into three alpha particles, while substantial, has very few of the alphas passing through the tight collimation slits in the energy selector location. The proton gantries seen in figure will not transport the much stiffer carbon beam, but at least one treatment room can be outfitted with the proton gantry for those patients receiving proton treatments from this cyclotron. In the following section, the issues associated with gantries suitable for carbon beams.

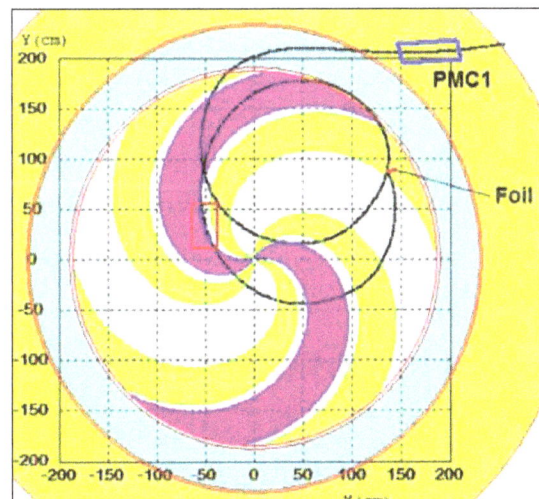

Proton extraction from the C400. The stripper foil reduces H_2^+ to two protons that spiral in the azimuthally inhomogeneous field to the extraction point at the top of the figure.

Superconducting Technology in Isotope Cyclotrons

Radioisotopes have been used for many years as tracers. An amusing anecdote concerning the first recorded use — in 1911, long predating accelerator production — is told about the young Hungarian chemist (and future Nobel Prize winner for his work on tracers) George de Hevesy, who, during on a stay in England at Ernest Rutherford's laboratory, spiked his dinner with radium to confirm that his boarding house matron was using leftovers instead of fresh food. But, with the advent of artificially produced isotopes, radioactive isotopes came into widespread use both as diagnostic agents in medicine and industry (isotopes emitting low-energy gammas), and for therapeutic purposes (with alpha or long-range beta emitters).

For medical applications, a key consideration was the isotope lifetime, and the distribution network to deliver the isotope to the end-user without excessive decay losses. Isotopes with several-day half-lives could be produced in concentrated centers and distributed to use sites. The workhorse of this production center became the ~30MeV cyclotron; IBA's Cyclone ~30 is a prime example.

PET scanning uses positron emitters, which typically have short half-lives — too short to permit the delivery time to be more than a few hours. For this application, having production capability as close to the use site as possible is very important; most desirable is being directly in the hospital or research center where the clinical studies are performed. Many small cyclotrons are today commercially available: compact, self-shielded 10–15MeV machines, with normal-conducting magnets and automated systems that cycle the target material (usually in liquid or gas form) through the accelerator beam to autochemistry systems to produce a labeled pharmaceutical ready for clinical administration. The largest market is for ^{18}F (110min half-life) attached to deoxyglucose, a metabolic fuel that is selectively absorbed from the bloodstream in areas exhibiting high glucose uptake, such as cancer cells. But other light isotopes are also of interest: 11C (20 min $T^{1/2}$), 13N (10min $T^{1/2}$) and 15O (2min $T^{1/2}$), probing different metabolic processes and targeting specific sites and functions.

These systems, provided by IBA, Sumitomo, ACSI or other commercial suppliers, accelerate H⁻ ions and extract via stripping. The stripped ion, now a proton, bends in the opposite direction in the magnetic field, and so is easily and efficiently extracted. This technique has vastly improved the usefulness of small cyclotrons: extraction of positive ions from a cyclotron is difficult and lossy, and performance has been limited by heat and activation in the extraction region. H⁻ extraction is essentially 100% efficient and avoids all these issues. Another advantage is that beam can be shared between two strippers, splitting the heat load on foil and target (quite often, target heating is the limiting factor), and essentially doubling the production rate.

Though compact, the shielding clamshell for these self-contained cyclotrons is typically 3–4m across. There is definitely room for a more compact solution.

Oxford Instruments' OSCAR Superconducting Cyclotron

The first superconducting isotope cyclotron was OSCAR, developed by Oxford Instruments in 1989. This unit, shown in figure, was an amazingly sophisticated instrument. It produced external beams of about 150 μa at 12MeV. Superconducting coils provided an average field of 2.36 T, with no return yoke. A thin steel sleeve surrounded the coils, with a set of superconducting bucking coils to channel flux through the steel sleeve, and provided essentially complete cancelation of any stray field, thus accounting for the very low weight of the total system — less than 2 tons. All the superconducting coils were run in persistent mode. Cooling was provided by a liquid helium bath, but a cryocooled 20K heat shield kept helium boiloff to where refilling was needed only once every two months.

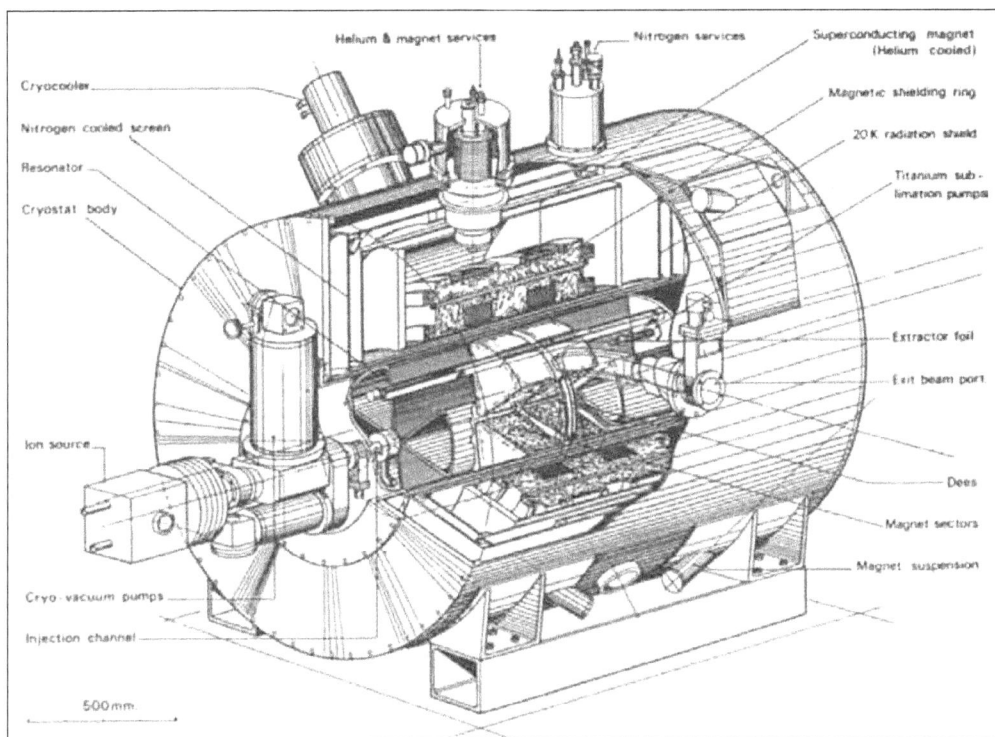

Oxford Instruments' OSCAR 12MeV superconducting yokeless isochronous cyclotron.

Three sets of sector steel poles in the 500-mmdiameter warm bore of the magnet provided the vertical focusing, and isochronicity; the valleys were open to allow full room for the 108MHz resonators acting as terminated transmission lines. H⁻ ions were produced from an external source, easily accessible for maintenance, and axially injected into the plane of the cyclotron. A stripper, at 220mm radius, provided extracted proton beams with high efficiency.

A total of nine production units were known to have been built. Today, several are still in operation.

Ionetix Isotope Production Cyclotrons

Ionetix has recently demonstrated a compact highfield (6 T) H⁺ cyclotron — named the Isotron — for ammonia ^{13}N (10min $T_{1/2}$) production, with protons at 12.5MeV striking an internal target. Ammonia ^{13}N is a superior cardiac diagnostic agent but has not yet been used outside of research hospitals.

This extremely compact Nb3Sn-based system has a mass of less than one ton and consumes less than 7 kW of power during operation. It is optimized to produce unit doses of the ammonia tracer on demand. Two unit-dose systems are presently under construction for deployment at leading medical centers in 2012 and it is expected that commercial sales will begin in 2013.

Beam Delivery and Compact Gantries

Gantry Considerations

Implementing the requirement for isocentric delivery drives the size of the room in which the patient is treated. Proton gantries, shown schematically and photographically in figures, are usually on the order of 12m in diameter, and 10m long. Considering that the entire room must be adequately shielded — 250MeV protons require about 2m of concrete — the cost for enclosing a large gantry is very high. As the roof must span the entire width, too, structural integrity requirements add to the cost.

The mechanical structure of the gantry is also a considerable expense. The requirement for precision of the beam at the isocenter, often expressed as a "circle of confusion," is on the order of a millimetre radius. That is to say, the on-axis beam must pass through this millimeter sphere regardless of the orientation angle of the gantry. This places severe limits on allowed deflection of the structure, and considering the weight of the magnets requires a massive, rigid structure. The most common proton gantries weigh about 100 tons.

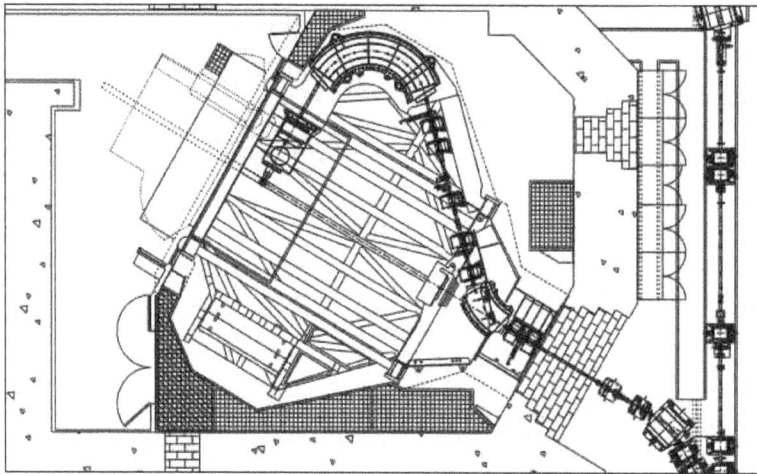

Plan view schematic of the IBA proton gantry. To keep the magnet size and weight down, the beam-spreading system is placed after the last magnet, but this does increase the swept radius of the gantry because of the need for drift distance for the beam to reach the required field size.

Considerable cost savings could result from developing an effective "compact" gantry. Various beam line configurations have been considered, and in some cases constructed, for reducing the gantry diameters such as the off-center configuration where the patient couch and 90° magnet rotate about a common center. While saving space, medical personnel have not been enthusiastic about this design because of inconvenient patient access. Nonorthogonal configurations — beam swept around a cone of, say, 60° instead of the full vertical plane — also save space, but lead to treatment configuration compromises that detract from their usefulness (for one thing, no vertical beam).

Clear are the advantages of increasing the field strength in the last magnet, achievable with superconductivity — not only in decreasing the bending radius with higher fields, but also in decreasing the overall weight of the transport system and so reducing the demands on the structural elements of the gantry. However, the problem is far from simple.

Photograph of the IBA gantry support structure. The heavy, 100-ton structure is needed to preserve stiffness and alignment accuracy of beam through all rotation angles.

Gantry size is not solely driven by magnet strength. Upon emerging from the last magnet, and directed toward the patient, the beam must be spread laterally to cover the full size of the field in both transverse planes. If the aperture of the last bending magnet is small, then the spreading system must introduce the required angular divergence, and the drift distance from magnet to patient must be large enough to achieve the largest field size specified. The spreading system (along with field-definition and beam-monitoring instrumentation), sometimes referred to as the beam "nozzle," plus the required drift distance could add as much as 3m to the radius of the gantry.

The gantry diameter can be reduced by placing at least part of the beam-spreading system upstream of the last bend, which increases the effective drift distance. This has the advantage, as well, of making the beam more parallel, and increasing the SAD (source–axis distance). However, the cost of this is that the aperture of the last magnet must be large enough to accommodate the spreading beam. Again, superconductivity can make this requirement much less onerous.

One problem that must be addressed, though, is energy variability. Each treatment involves particles with a wide range of energy, corresponding to the depth and thickness of the tumor in the beam direction. For instance, a tumor 10 cm thick located between 10 and 20 cm inside the body will require protons of 120MeV at the front (proximal) edge, and 175MeV at the back (distal) edge (not counting energy loss in the spreading system); 34% and 83% of the maximum rigidity, respectively. As energy selection is done shortly after the accelerator, the entire beam line must accommodate the range of rigidities. And, to meet the treatment time requirement, the variation in magnet settings must be accomplished in no more than a few seconds. Ramping superconducting magnets at this rapid rate will present a considerable challenge. As we will see, various approaches are being taken to address this issue.

Observe that the simplest technique, namely running the full energy beam through the gantry and performing the energy degrading just after the last magnet, is possible but carries the large drawback of substantially increasing the neutron flux the patient is exposed to from the inevitable loss of particles in this degrading system. Studies are being performed of the effect of this whole-body neutron exposure at proton facilities that use passive scattering, and where heavy collimators close to the patient are used for lateral field shaping.

Proton Gantries

In essentially all commercially supplied proton gantries today, the beam-shaping system, whether scanning or passive scattering, is located after the last magnet. This adds as much as 6m to gantry diameter. For 2.5 T-m rigidity, and the 2T magnet, the proton orbits have a radius of 1.25m, contributing 2.5m to the diameter. Steel yoke and gantry support structure can add another 2–3m, making it difficult to design a gantry smaller than 11 or 12m in diameter. Use of superconducting magnets, with 4T or higher fields, would save 1.25m from beam bending, and the smaller magnets might save another 1–2m in the gantry diameter, bringing a superconducting proton gantry to about 8 or 9m in diameter.

ProNova Solutions/ProCure

Sketched in figure is the concept proposed by Ionex and ProCure, and being developed by ProNova Solutions for a light proton gantry based on superconducting magnets. The configuration, using combined function (dipole–quadrupole) windings into an achromatic configuration, and magnet apertures allowing up to about a 9% momentum acceptance with high transmission efficiency.

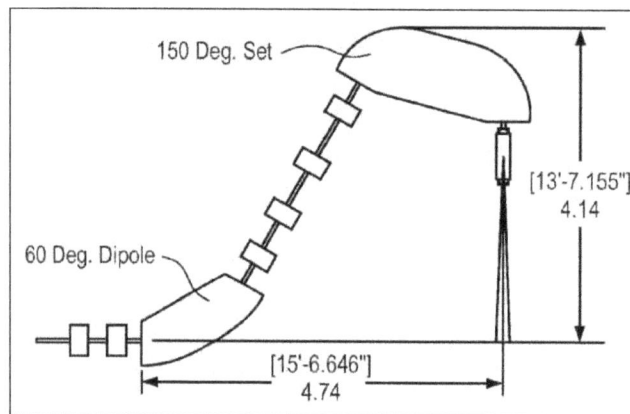

Proton gantry being developed by ProNova Solutions and ProCure. Based on sets of achromatic superconducting magnets, even with the scanner mounted in line with the patient, this gantry still offers size and particularly weight advantages over conventional proton gantries.

The beam scanning system is located in line with the patient, and the gantry diameter is a bit over 8m. The total mass of magnets is expected to be less than 5 tons, which will substantially decrease the total weight of the gantry. Further details are not yet available, but the goal is to have a system in the field in 2015.

Ion Beam Gantries

Where superconductivity can have the largest impact is on carbon beam delivery systems. All facilities treating patients today with carbon ions utilize fixed beam orientations — horizontal, vertical or oblique (45°). While robotic patient-positioning systems can provide some flexibility in beam port orientation, the clinical world still calls for isocentric delivery.

The first and only existing carbon gantry has been built at Heidelberg. This gantry, weighing about 600 tons, has scanning magnets upstream of the last bending dipole, which as a consequence has a bore of 20 cm and a gap of 20 cm. But this does reduce the distance between the end of the magnet and the patient. Still, the diameter of the gantry is 13 m, and its overall length is 25 m. This gantry has just completed its commissioning and treated its first patient in November 2012.

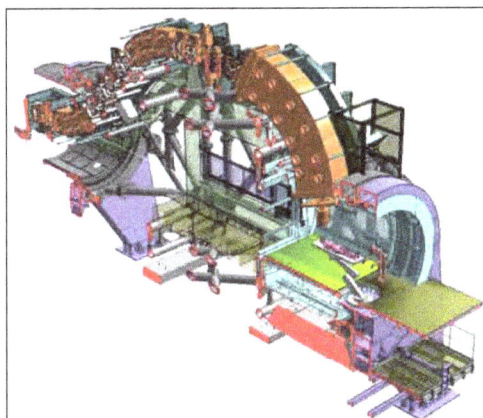

Carbon gantry at the Heidelberg ion therapy center. The massive steel structure supporting the magnets has been cut away in this schematic. Scanning magnets are located before the last 90° bend. Compare the size of the patient on the couch with the cross section of the last magnet.

NIRS/HIMAC

After the closure of the Bevalac, the torch for ion beam therapy was picked up by HIMAC at the National Institute for Radiological Studies in Chiba, Japan. This large complex, with two synchrotrons and three treatment rooms, started treating patients in 1994. The treatment rooms are configured with static ports — one room with a horizontal beam, one with vertical delivery and the third with both a horizontal and a vertical port.

A new project, initiated in 2006, will build a second therapy complex with three treatment rooms— two with both horizontal and vertical beam lines, and the third with an isocentric gantry.

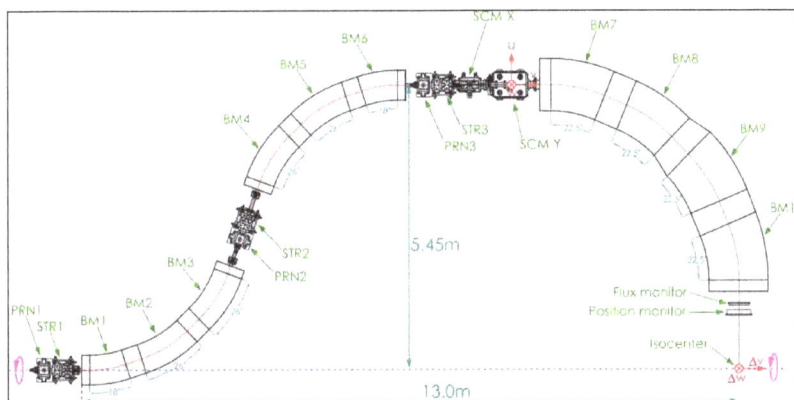

Layout of the HIMAC gantry, showing the two sets of small-bore (60mm dia.) magnets BM1-6 and the last 90° set with the maximum bore diameter of 290mm.

This gantry, schematically shown in figure and 30, also places the scanning magnets before the last 90° bend, requiring large apertures in the last magnets. The overall dimensions of the volume swept by the gantry are 5.5m radius and 13m length. This makes it approximately the size of a current proton gantry, and it is estimated that the weight of this gantry will also approximate that of the current proton installations.

The first two sets of magnets, labeled BM1-6 in figure, have the same cross section and bending radius, with two sets of 70° bends (18°, 26° and 26° for BM1,2,3, and inverse for BM4,5,6) and a 30mm bore radius. The last four magnets each provide a 22.5° bend, and have bore radius values of 85, 120, and the last two 145mm to accommodate the spreading beam. The field size at the iso-center is 200 × 200mm.

The maximum dipole field in the small-bore magnets is 2.88T; in the large-bore magnets, 2.37T. Operation is planned to coordinate with the recently developed multienergy flat-top mode for extracted beam from the synchrotrons; the small energy changes between slices can be accommodated by the superconducting magnets in about 200 milliseconds, to change current and allow for settling, then a few hundred milliseconds to deliver beam to that slice through the scanning system.

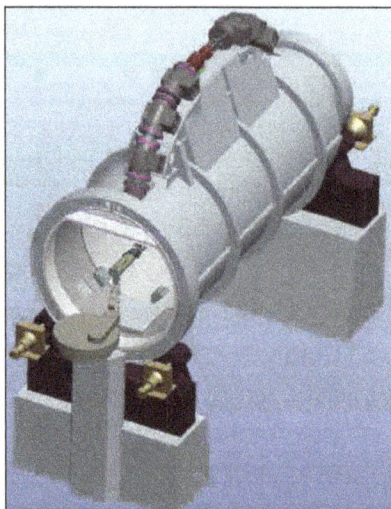

Schematic of the new HIMAC superconducting gantry.

All magnets are "combined-function," with two sets of NbTi coils, the inner 8 layers wound in a $\cos(2\theta)$ configuration to provide the quadrupole focusing field, the outer 26 layers in a $\cos(\theta)$ configuration to provide the dipole field. Figure shows the profile of this magnet, with the warm beam tube, a superinsulation layer to the stainless coil base, coils and the cold steel yoke. The total diameter of the cold mass is 500mm. Figure 32 shows the two magnets (BM2 and BM10) that have been built and are under test.

CEA/ETOILE

The ETOILE project, aimed at carbon/proton therapy in Lyon, France, has been under study for about 10 years. A recent effort has been a joint CEA (Saclay)/IBA study for the design of a gantry; the concept is for a "classical" Pavlovic layout, with normal magnets except for the last 90° dipole. For a radius of curvature of 2 m, the maximum bending field will be 3.2T, while the highest field in the coils will be 5.4T, comfortable for the NbTi conductor. The right image in figure shows the 12 planar coils, which include 6 coils for active shielding. With no iron, the weight is substantially reduced. The planned slew rate (dB/dtmax) is 0.04T/s, adequate to treat one slice of a carbon field in about 1.5 s.

Cooling is planned with no liquid reservoir; the 10 Sumitomo cryocoolers, capable of operating at any orientation (necessary to accommodate gantry rotation), provide adequate cooling, though the cooldown time is long (about 20 h for quench recovery from 70 K, 1 week from T_{room}).

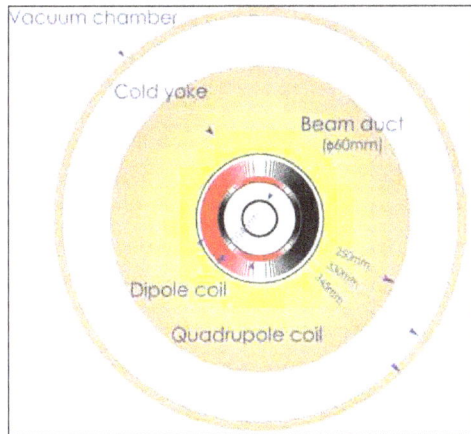

Cross section of the small-bore magnet. Cold steel encases the coils.

The study has concluded that the concept is entirely feasible, and the project is awaiting funds for further development of the design, and for building of prototypes.

Coil packages and finished magnets for the small-bore (BM2) and large-bore (BM10) combined-function HIMAC gantry magnets. The length of the magnets is about 1m in both cases, the outer diameter of BM2 is 690mm, and BM10 1.24 m.

Tilted Double-helix Coil

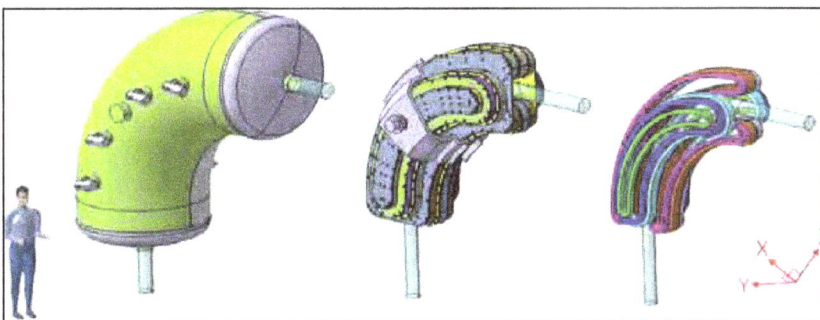

CEA/ETOILE large-bore 90° bend dipole. The left most image shows the outer vacuum vessel with cryocoolers; the other two images show the coil packages, plus bucking coils for active shielding, saving weight of steel.

Also aiming to produce a more compact and lighter gantry, a team at LBNL is developing a novel magnet design for the final 90° large-aperture toroidal bend based on tilted double-helix (also referred to as canted cosine theta, CCT) windings. Proposed in 1970 by Meyer and Flasck, a dipole field can be generated by overlaying solenoidal coils that are tilted in opposite directions. This configuration cancels the solenoidal field and sums the dipole component of both. In addition, quadrupole and higher-order fields can be obtained with appropriate winding schemes, as shown for the quadrupole case in figure. Thus, with appropriate windings, a combined-function magnet, with appropriate corrections, can be produced in a compact, lightweight package.

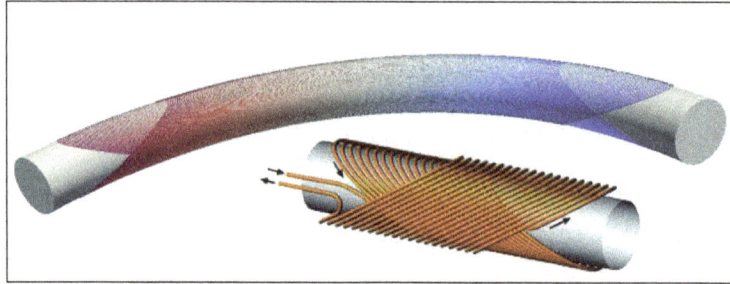

Double-helix canted coil windings showing the current path. The solenoid component is canceled, while the dipole components add. The configuration can be easily wound on a toroidal mandrel to make effective gantry magnets.

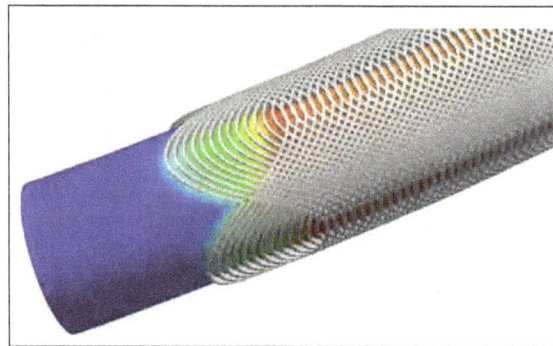

Tilted coil configuration for a pure quadrupole field.

In fact, Goodzeit et al. have shown that the multipole components can be introduced into a single double-helix winding by appropriately shaping the path of each winding, as illustrated in figure.

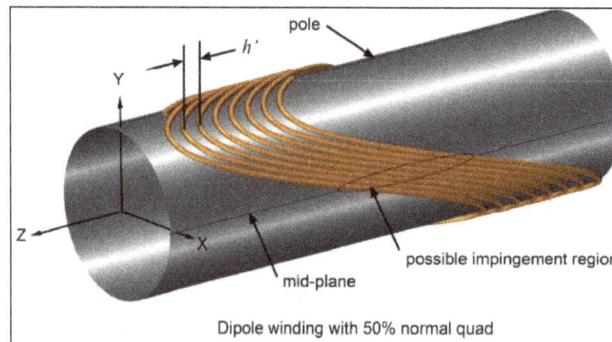

Combined-function configuration proposed by Goodzeit et al., obtained by varying the spacing of each winding around the circumference of the mandrel.

The LBNL group has calculated the effect on the field of windings along a toroidal path, as would be needed for the last 90° gantry bend, and determined that field errors can be corrected with appropriate sextupole compensation. Small prototypes for both dipole and quadrupole magnets have been built with NbTi, and tested at LBNL. The design simplicity and low construction costs give confidence that this technique can find very effective application in compact gantries.

FFAG

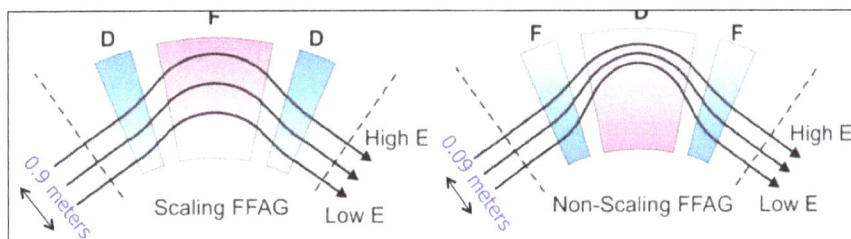

Schematic of FFAG "scaling" and "nonscaling" configurations. A "scaling" lattice preserves the same tune for different momenta, not important for single-pass transport lines. The configuration can be designed to maximize momentum acceptance in a minimum aperture. FFAG basically offers excellent strong-focusing lattices.

Fixed field alternating gradient (FFAG) lattices have been proposed for many different applications; their particular advantages are strong focusing and very high momentum acceptance without having to change the magnet settings. Figure illustrates the basic concepts for scaling and nonscaling configurations, showing alternating inward and outward bending dipoles, with also alternating gradients corresponding to each dipole direction. For a simple transport line (as opposed to an accelerator), transit times and orbit tunes are unimportant, and the lattice parameters can be optimized for maximum momentum acceptance in a minimum aperture. Trbojevic has designed FFAG gantries suitable for carbon beams with momentum acceptance δp/p of ±20%, so a single field setting of the gantry magnets could transport carbon beams between 200 and 400MeV/amu. This acceptance covers carbon ranges from about 9 cm to over 25 cm, quite adequate for most treatments. Note that the outward-bending elements of the lattice will increase the radius of the gantry, but strong superconducting magnets can compensate for this; Trbojevic's gantry designs have radii of about 4.5m, with magnet apertures of around 4 cm. The magnets are challenging, but concepts are believed possible using winding configurations capable of creating all the necessary field configurations, producing the required fields and gradients, and still remaining within the technical limits for the NbTi conductor.

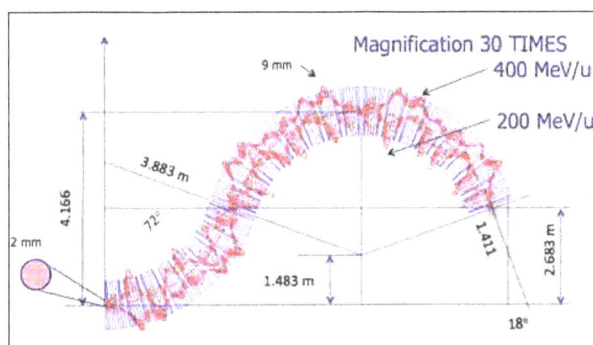

Nonscaling FFAG gantry for carbon ions, showing trajectories for 400MeV/amu and 200MeV/amu. Both sets easily fit inside a 4-cm-diameter bore tube. The radius of the full gantry is no more than 4.5 m. Following the exit shown, scanning magnets and a special FFAG cell will be designed to bring the beam to the perpendicular orientation and allow for the required 20 × 20 cm scanned field.

MRI Magnets

Advances in Imaging

Imaging quality has made huge strides in recent times. X-ray CT (computerized tomography) scanning, MRI (magnetic resonance imaging) and PET (positron emission tomography) are providing detailed spatial and functional information that has totally revolutionized diagnostic capabilities and is in addition enabling pinpoint precision for target definition in radiation therapy. Each has particular strengths, and the greatest diagnostic value is being obtained by combining information from all three. The greatest challenge comes in overlaying or matching the images from each, known as "fusion," and is now being addressed by combining modalities into the same apparatus; so, for instance, a patient is scanned with CT, and in the same setup a PET scan is performed, thus assuring alignment of the coordinate systems for both diagnostic procedures. PET–CT devices are now available, and PET–MRI is close to commercial reality. Software fusion of CT and MRI images is quite well developed.

Computerized Tomography

CT reached commercial viability in the 1970s. It relies on passing a thin beam of X-rays (typically 80 kVPh) through the object being imaged, measuring the attenuation of the X-ray beam in a detector beyond the object. This beam is rotated through 360° along with the detector, collecting attenuation information at each angle. The full data set is passed through a back-projection reconstruction algorithm to produce an image of tissue density in the plane of the slice. The patient is moved to scan subsequent slices, collecting data to provide a full three-dimensional image of the region of interest. Figure shows a representative scan with a modern CT instrument. By using fan beams and arrays of detectors, and more recently spiral motion of the rotating mechanism, the efficiency of data collection has significantly improved, reducing scanning times to a few seconds. Image quality is also being enhanced, by using dual-energy scan systems (e.g. 80 and 140kVP). Image resolution is determined by the size of the X-ray focal spot and the detector size, with pixel sizes now in the submillimeter range.

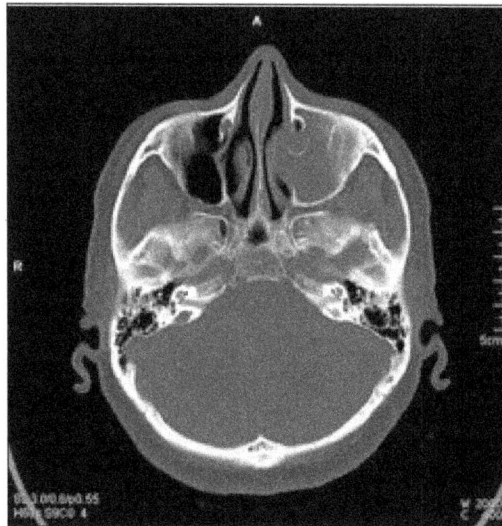

CT image showing high contrast between bone (skull) and soft tissue, but differentiation of soft tissue structure is difficult.

Attenuation at this energy is dominated by Compton scattering off electrons in the tissue, which in turn is highly dependent on atomic mass; so bone, with a large amount of calcium, is more easily imaged than soft tissue. The reconstruction algorithm assigns to each pixel a "Hounsfield number" related to the electron density in that pixel. As electron density is closely correlated with energy loss or attenuation of the beam, CT is the primary tool for treatment planning for therapy with external beam radiation.

Positron Emission Tomography

PET also relies on detection of photons, but in this case the photons are 511 keV back-to-back gamma rays emitted when a positron annihilates with an electron. The positrons are emitted by a radioactive nucleus disintegrating inside the subject; these positrons slow down and encounter an electron typically within a 1–2mm distance from the parent nucleus. The annihilation gammas are given off almost perfectly aligned, so detecting the two photons allows one to draw a line defining the path along which the decaying nucleus will be located. By observing many such coincident pairs, a density map of the radioactivity can be generated.

The radioactive material is attached to a substance that is selectively absorbed in the tissue being imaged, so the PET spectrum will identify these tissues. A common system is to attach 18F to deoxyglucose, which is absorbed from the blood to fuel metabolic activity. PET can be used for studies of activity in the brain, or inversely areas of the brain that have been inactivated by disease such as Alzheimer's.

Identification of metastatic lesions, also characterized by higher metabolic activity, makes PET of extreme value for locating targets for therapy. Figure shows a whole-body PET scan, identifying locations of metastatic disease, secondary tumors located far from the primary mass.

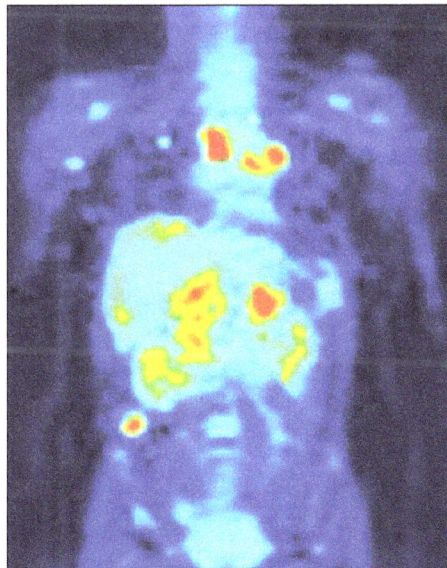

Whole-body PET scan showing metastatic tumors, identified by high metabolic activity and the uptake of FDG (^{18}F-deoxyglucose).

The production of the short-lived positron emitters is a major application for small accelerators. Desirable half-lives for these isotopes should be short enough to render insignificant the dose of radiation to the patient following the procedure, and so are less than an hour or two. Consequently, isotopes that are not available from a "generator" should be produced very close to the use point to minimize the loss of activity due to long delivery times. Large PET centers tend to have their own cyclotrons, with automated chemistry to transport and process the target material directly into the pharmaceutical to be used for the procedure.

This places great emphasis on development of accelerator and chemistry systems that are compact, reliable and inexpensive.

Magnetic Resonance Imaging

MRI became a practical diagnostic tool in the late 1980s, and has overtaken CT for image resolution and quality. Furthermore, since it does not involve ionizing radiation, the not-inconsequential dose to the patient from high-quality CT scans can be avoided.

The subject is placed in a strong, uniform magnetic field, and orthogonal gradient coils adjust the field at each location within the scanned volume to bring hydrogen nuclei (protons) into resonance with an applied radio-frequency signal. A complex series of RF pulses and sequence of time variation for the gradient coils, and a sensitive pickup antenna, detect relaxation of the proton spin orientation in the magnetic field at each {x, y, z} voxel coordinate to produce exquisitely sharp images of the anatomical structures. A combination of proton density and different relaxation times in different tissues can be used to adjust contrasts, and tease out very detailed information on tissue types.

The "traditional" MRI system today uses either 1.5T or 3.0T solenoid fields, with superconducting coils operating in persistent mode. Development is proceeding on systems at both higher and lower fields, for different clinical and research objectives.

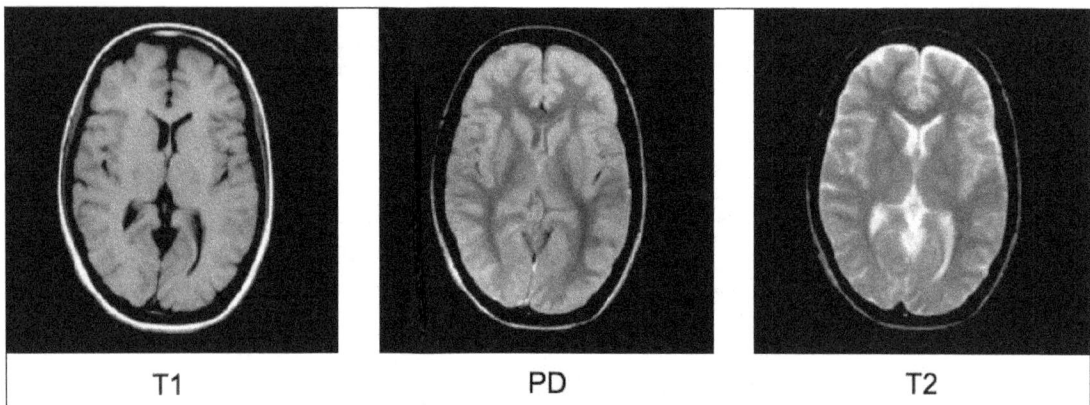

| T1 | PD | T2 |

MRI images of the brain, with different weightings, corresponding to different timing sequences in the pulse train. PD is proton density, a basic map of hydrogen distribution. T1 and T2 reflect different relaxation and dephasing times. Tissues affect these relaxation times in different ways, so for instance cerebrospinal fluid has a much longer relaxation time and is highlighted in the T2 image.

While image quality and information for diagnostic purposes is excellent, absolute coordinate accuracy is more difficult to establish than in the case of CT scans. The coordinates of the voxel being examined are obtained from the magnetic field map summing the main field and the instantaneous settings for the three gradient coils, and any effect that can distort the magnetic field at the location of the resonating nucleus will result in errors in coordinate location. The currents in the gradient coils predict the field at a voxel to be a given value, B(x, y, z), but a chemical shift (caused by screening of the field at the nucleus due to orbital electrons or molecular orbitals), paramagnetic susceptibility of tissues (such as iron in blood), or any magnetic material in close proximity to the scanned volume can introduce a ΔB that will make the nucleus resonate at a slightly different frequency, and so appear to be displaced from its actual site according to $\delta B/\delta r$ of the gradient coil configurations. For diagnostic purposes this is usually of little consequence, but it is important for treatment planning with external radiation beams. MRI and CT images are fused to compensate for this: the fused image displays soft tissue which cannot be seen in CT, while the more accurate dimensional information from CT is used to more reliably fix the target coordinates.

Historical Development of MRI

The first whole-body scan was performed by R. Damadian in 1977 in the Downstate Medical School, State University of New York in Brooklyn. The superconducting magnet with a 53 in. (1.4 m) warm bore was built in the Medical School shop, using NbTi wire. The coil was bathed in liquid helium, with two separate (concentric) 77K toroidal heat shields. Designed to run at 0.5 T, most of the R&D for their MRI imaging was performed at 0.1T. While the resolution was quite poor, it did demonstrate the viability of imaging with nuclear resonance. Though this was the first MRI image obtained, techniques developed by P. Lauterbur and P. Mansfield were credited with showing the path to the technology in use today. While eyebrows were raised, Damadian was passed over when the 2003 Nobel Prize in Physiology or Medicine was awarded to Lauterbur and Mansfield of laying the foundation of modern-day MRI.

R. Damadian, L. Minkoff, M. Goldsmith and the "Indomitable," the world's first MRI scanner which they developed.

Initial developments toward commercialization of MRI were focused on relatively-low-field, mainly resistive magnets, of 0.5T or below. However, in 1980, Oxford Instruments convinced GE that it could provide a superconducting solenoid with a bore large enough for a patient to fit in (~65 cm), with a field of 2T. The first magnet delivered to GE at its Niskayuna, New York research laboratory exhibited a substantial downward drift in the persistent current at high fields, but at 1.5T it did run stably and reliably. The magnet was delivered in 1982, and the GE team began assembling a system capable of imaging. Jumping to this high field was considered at the time a high risk, because it was believed that the RF for proton resonance at 1.5T (63MHz) would be attenuated in the body and so insufficient signals would be obtained. Fortunately, this turned out not to be the case, and the first brain image at 1.5T was obtained in 1983.

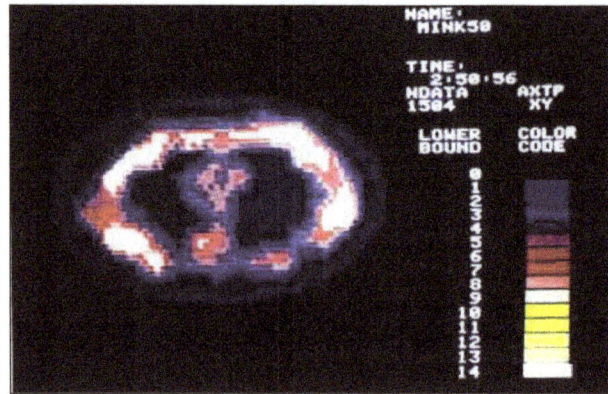

The first MRI image of human anatomy, obtained with the "Indomitable" device: L. Minkoff's chest.

The success of this endeavor led to rapid development and deployment of MRI systems, with the 1.5T magnetic field becoming the de facto standard for these clinical instruments. Figure shows a modern clinical instrument; the bore is of the order of 70 cm, and the length of the magnet from cover to cover is about 1.5 m. The units weigh several tons. Initially cryostats contained up to 1000L of liquid helium, but the development of effective cryocoolers by Sumitomo Heavy Industries in the early 1990s enabled substantial cost reductions by eliminating the need for liquefaction plants and the large inventory of helium.

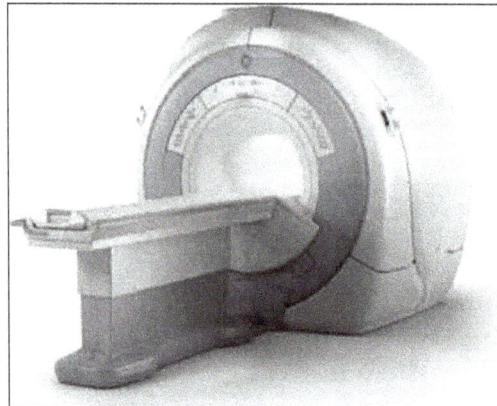

Typical MRI scanner in clinical use: a GE Brivo MR355 1.5T scanner.

In the early 2000s, scanners operating at fields of 3T were available for clinical use, with the appearance and specifications essentially identical to those of the 1.5T units except for higher resolution, greater flexibility of imaging options and, of course, higher costs.

Safety Considerations

While MRI scanners are free of the ionizing radiation associated with CT and PET, other hazards related to the very large magnetic field must be considered.

Ballistic Hazards

Any magnetic object — such as a key chain, screwdriver, metal cart or other object — accidentally brought into the room containing the scanner could become a projectile capable of doing very considerable damage to the equipment, or should a patient be in the scanner, causing possibly life-threatening injuries. Mitigation of this requires detailed logging of all materials in the scanner room, and very tight discipline for personnel entering the room, especially regarding the contents of their pockets. As the magnets are run in persistent mode, i.e. are energized around the clock, this discipline extends to the off-hours janitorial crew and the mops and pails they use to clean the room.

To minimize the stray fields, and help with mitigating this projectile problem, manufacturers now provide magnetic shielding, either in the form of iron often built into the room because of the large bulk required, or through active compensating coils — separate superconducting windings that buck the fringe field.

Patient Screening

The high magnetic fields can have disastrous effects on individual's unfortunate enough to not have been carefully screened prior to placement in the MRI magnet. Some implants, such as pace-makers, cardiovascular catheters and surgical clips, may be severely affected by the fields.

One very serious consideration that must be explored during screening is whether the patient has received injury from an explosion or impact that may have left (magnetic) metallic fragments imbedded in tissue, organs or, worse, the brain or eye. The process of inserting the patient into the scanner is likely to cause these objects to move, with possibly serious injury to the patient. At the very least, severe distortions of the image will result.

Operational Housekeeping

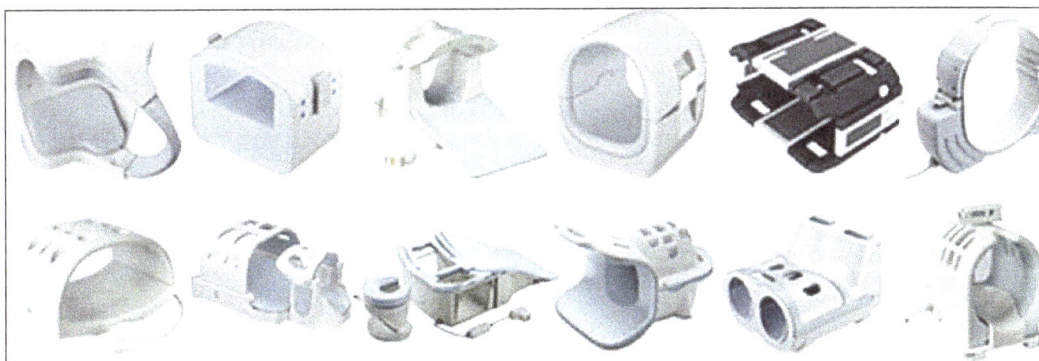

Collection of specialized RF coils tailored to fit closely to the anatomical site being scanned.

Obtaining good images requires having very strong and efficient RF transmitters and sensitive pickups. It is often not enough to use coils located at the edge of the inner bore, which would be used for a whole body scan. To obtain the highest resolution and contrast for a particular anatomical site, very specialized coils are fabricated that are placed in close proximity to the site, such as wrapping around a knee or elbow, or close to the skin for certain sites or organs. After the patient is placed on the couch, the coil is put at the appropriate location and the patient is then translated into the magnet bore. Great care must be taken by the technician that the cables to the RF coil are dressed properly along the patient, and specifically that there are no loops in the cables. Cases of severe burns have been recorded where inductive energy absorbed by such loops caused rapid heating of the cable. While already a problem with the 1.5T magnets, the effects become considerably more serious in the new research scanners operating at 7T, or even up to 12T.

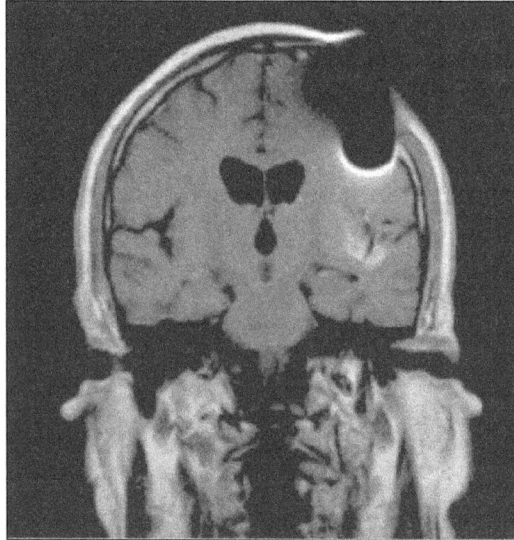

Distortion caused by a metallic fragment, in this case a small piece of shrapnel from a grenade explosion. The magnetic fragment has affected the magnetic field in its vicinity, producing an uncorrectable artefact.

Psychological Impact

The environment inside the scanner is not exactly the most pleasant for many patients, in particular those with tendencies toward claustrophobia. Although the technology is improving, scan times are still relatively long (several minutes at least), and the pulsing sequence of the gradient coils and RF system produces substantial acoustic noise. Tolerance to this environment should be explored during the screening process, and patients likely to experience problems discouraged from undergoing an MRI procedure.

Manufacturers are addressing this particular issue by technological improvements that decrease scanning time, reduce acoustic noise and prevent the patient from feeling like they are being stuffed into a torpedo-launching tube. Specifically, Toshiba claims to have improved the structural containment of the gradient coils to produce significantly quieter scans.

Research Directions

R&D is proceeding in two directions: toward higher fields for better resolution and faster scan times, and toward lower fields for various clinical reasons.

Lower Fields

Though it is known that higher fields produce better results, significant effort is being devoted to development of scanners operating at lower fields. The reasons for this are mostly clinical, and the challenge for the industry is to improve image quality to acceptable levels at these low fields.

Two main reasons driving technology in this direction are: to increase the accessibility of MRI to those patients who cannot tolerate the high fields due to implants or other at-risk materials that cannot be separated from the patient; and to open up the area containing the magnetic volume to reduce claustrophobia or make MRI accessible to heavier patients. An unfortunate demographic trend toward very large individuals, particularly in North America, is excluding an ever-increasing fraction of the population from MRI procedures with industry-standard devices.

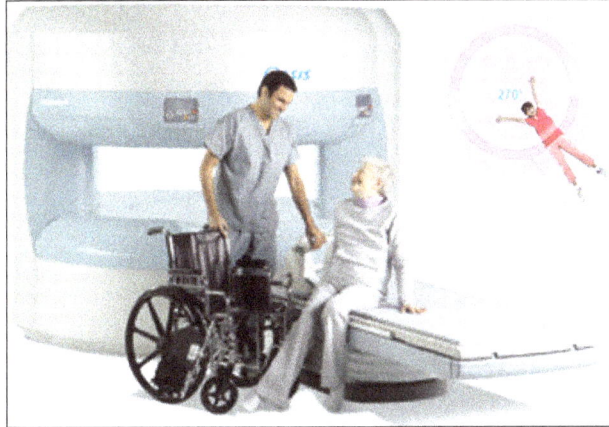

The Hitachi OASIS system, a split-coil vertical field 1.2T system with a very open configuration.

Higher Fields

Increasing the magnetic field offers advantages of substantial increase in signal strength, which in turn can be converted into very high spatial resolution and/or decreases in scanning times. Basic signal strength is related to polarization of the protons in the medium being scanned, i.e. to the number of protons contributing to the resonant signal. This is in turn given by a Boltzmann distribution of the form:

$$P = \frac{N^+ - N^-}{N^+ + N^-} = \tanh\left\{\frac{\gamma h B_0}{4\pi kT}\right\}$$

Where, P is the net polarization, γ the gyromagnetic ratio, h Planck's constant, B0 the applied magnetic field, and kT the Boltzmann thermal distribution factor. At room temperature in a 1.5T field, $P = 3 \times 10^{-6}$, quite a small number, but still adequate for generating a good signal. For very small values of x, tanh{x} expands as:

$$\tanh\{x\} \sim x - \frac{x^3}{3}$$

So polarization goes roughly linearly with the applied field B_0. By increasing the field from 1.5T to 3T, the polarization factor increases by a factor of 2; at 10T it is almost a factor of 7 higher. This will in itself have a huge effect on the quality of images produced, but the story does not stop there.

Image quality is related to the signal-to-noise (S/N) ratio. Noise factors can also be affected quite beneficially by an increased field, and detailed analyses estimate that overall improvement in image quality can be as high as $\Delta B2.5$.

Figure illustrates quite spectacularly the improvements in image resolution in increasing the B field from 1.5T to 7T. In addition, high-field magnets create the opportunity to do very elegant/exotic scans (for example, diffusion-weighted scans and others where one can actually look at the behavior of bundles of neurons) and spectroscopy that permits a determination of whether radiation/chemotherapy has actually controlled a tumor. Higher fields improve the sensitivity for fMRI (functional MRI) which measures brain activity by observing changes in associated blood flow.

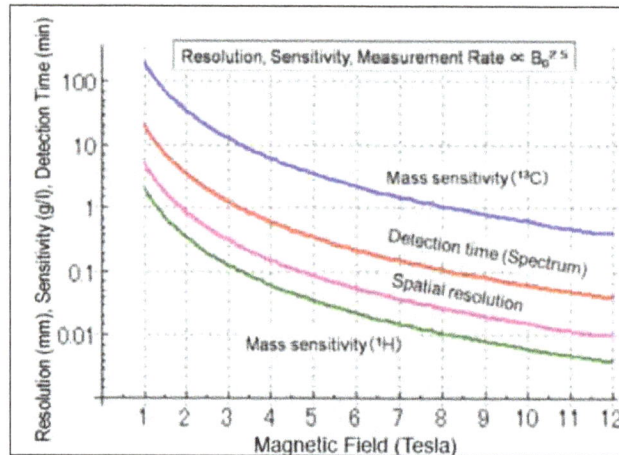

Estimated performance of MRI scanners for higher magnetic fields. Note the logarithmic vertical scale.

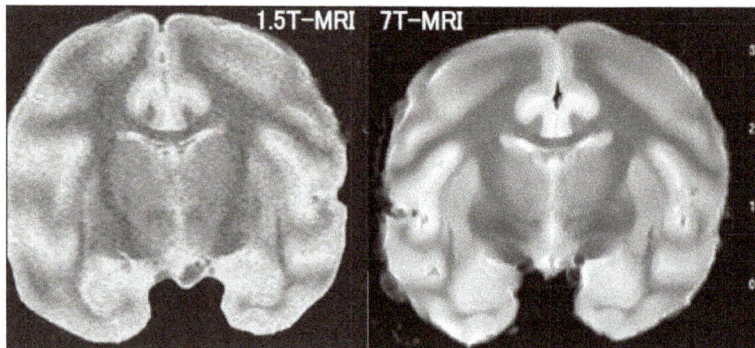

MRI images taken in 1.5T and 7T scanners. The improvement in image quality is quite apparent.

Improved resolution for spectroscopy research is seen in figure, which highlights the increased quality of chemical shift data at higher field strength. A chemical shift provides information on the molecular bonds of the resonating protons, and so can be an indicator of the structural environment. Increased signal strength also enables detection of otherwise weak signals from other chemical species such as 13C, also shown as a curve in figure.

For basic scans, improved S/N can be utilized in two different directions: increased resolution, as a smaller volume (voxel size) is needed to obtain data of adequate quality; or increased scanning speed, as the rate of data acquisition is higher. Figure shows both these effects. The respective curves are not independent: the same voxel size leads to faster scan speeds, or smaller voxels scanned for the same time.

A spatial resolution of tens of microns is viewed as possible in 10T fields, while at normal resolutions scan speeds can be high enough to capture normal biological motions such as the heart rate without the need for gating or sampling.

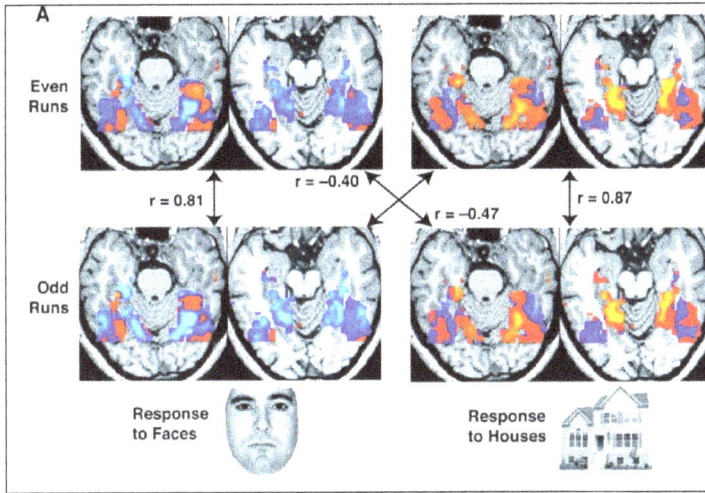

fMRI (functional MRI) image of brain activity responding to different stimuli.

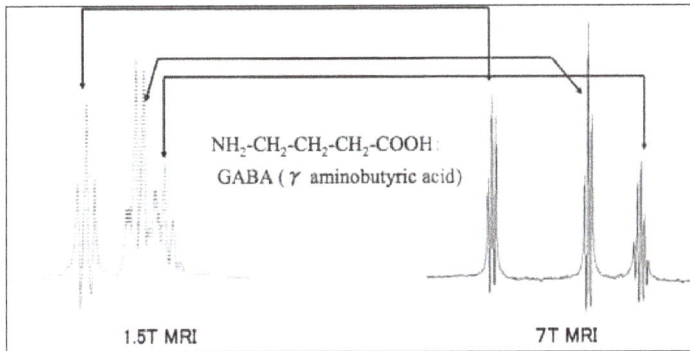

Spectra for proton resonance spectroscopy (MRS) for γ aminobutyric acid (GABA) at 1.5T
and 7T, also showing the significant improvement in the chemical shift data quality.

Table summarizes the magnet configurations in use today or contemplated for future use. At the highest fields, not only are the magnets exceedingly challenging, but adverse physiological effects of magnetic interactions with tissue or moving fluids (e.g. blood) could become important. In addition, the very high resonant frequencies will have attenuation lengths short enough that penetration of the RF signal into the subject will be a limiting factor in signal strength. For example, the attenuation length of 600MHz RF in tissue is about 7 cm, so imaging large objects in 14T fields will lose quality.

Table: Examples of MRI magnets in service or contemplated. Low-field magnets are usually configured with split coils to allow unrestricted access to the scan plane. 1.5T and 3 T systems are the workhorse units for clinical applications. Research units at 7T and 9.4T are operational today at several centers. The Iseult project is building an 11.7T, 90-cm-bore unit. Higher-field magnets are planned for animal studies, but will have smaller bores.

Field (T)	Proton frequency (MHz)	Bore size (cm)	Application
0.5	21	Open, split coil	Clinical
1.5	63	~70	Routine clinical
3	125	~70	Advanced clinical
7	300	~70	Human research, near-term clinical

9.4	400	~70	Research
11.7	500	90	Advanced research
14.1	600	40	Animal studies
17.4	750	25	Small animal studies

Magnets up to 9.4T are uniformly made with NbTi conductor; however, for fields of 11.7T and higher, critical fields and currents make this conductor problematical. The Iseult project utilizes NbTi, but requires cooling with superfluid He at 1.8K, substantially affecting the construction and operation cost of the unit. Nb_3Sn would be required to operate the 11.7T magnet at 4.2 K, or to reach higher fields.

High-field MRI Research Centers

Numerous research centers and projects are operating today. Two will be highlighted: the Center for Magnetic Resonance Research (CMRR) at the University of Minnesota, and Iseult/INUMAC at CEA Saclay.

CMRR

Specializing in research with high-field systems, this center operates numerous scanners, from a commercial 3T Siemens Trio unit to a 9.4 T, 65-cm-bore system provided by Magnex Scientific (now Agilent). This unit saw the first research with human patients at this field, published in 2006. Two 7T units (Magnex — passively shielded; and Siemens — actively shielded) and one 4T unit (Oxford), all with 90 cm bores, complete the instruments on hand. A research project with Agilent is underway to develop and install a 10.5T, passively shielded, 88-cm-bore system. This magnet will use NbTi cooled to 3K.

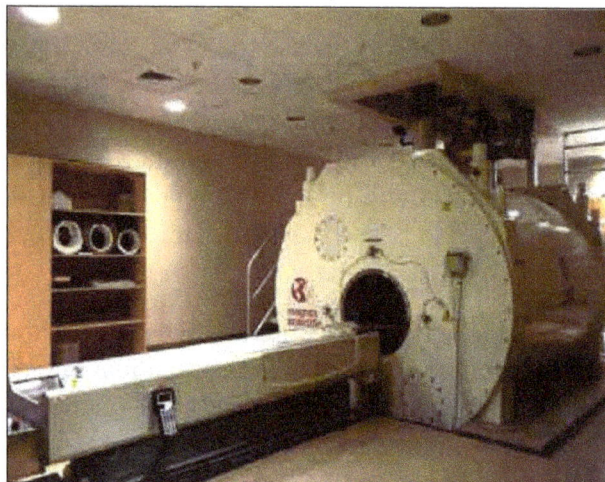

Photo of the 9.4T whole-body scanner at the University of Minnesota's Center for Magnetic Resonance Research, site of the first scan with a human subject at this high-magnetic-field value.

Iseult/INUMAC

The centerpiece for the French NeuroSpin center at Saclay will be an 11.7T whole-body scanner, which is at present under construction. Currently operating are a 3T "trio TIM" and a 7T research, whole-body scanner, both from Siemens (90-bore); a 7T "Pharmascan" unit (active shielding, 16 cm bore) and 17.2T "BioSpec 170/25" (active shielding, 25 cm bore) units from Bruker. The latter uses NbTi at 2K.

The 11.7T, 90-cm-bore scanner is being developed as a joint French–German project, under the leadership of Pierre Vedrine, CEA Saclay. NbTi (cryostable — flat Rutherford cable) in direct contact with supercooled (1.8K) helium is wound into 170 double-pancake coils of 1m inner diameter and 1.9m outer diameter, for a total solenoid length of 3.8m. Concentric shielding coils, outside the main solenoid, reduce the stray field to acceptable levels. Figure shows a cross section of the cryostat with solenoid and shielding windings, cutaway projection of the scanner. The current status is that the Technical Design Report has been completed, prototype tests have been concluded and construction contracts are underway. Delivery of the system, for commissioning, is scheduled for 2014.

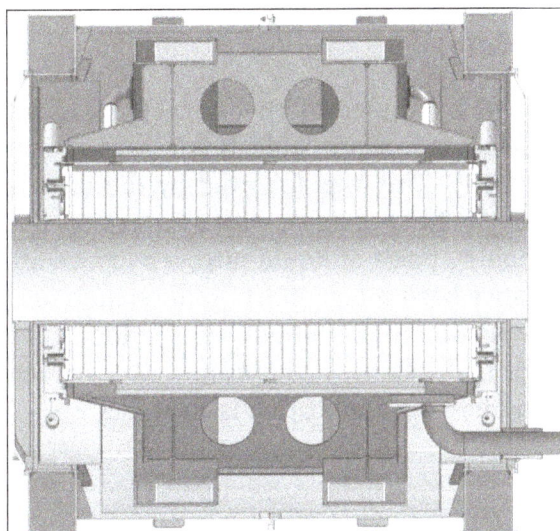

Cross section of the Iseult 11.7T cryostat, showing the main field pancake coils and the two (larger-radius) active shielding coils. The bore is 90 cm, and the total length of the warm tube is 3.8m.

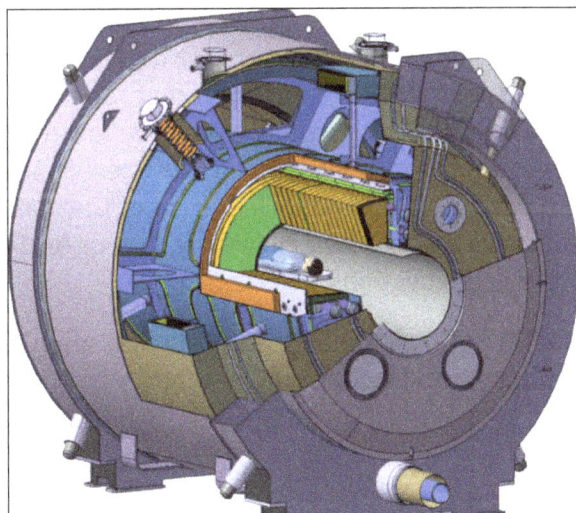

Cutaway of the Iseult 11.7T magnet and cryostat assembly.

Superconducting Transformers

The advent of high-temperature superconducting (HTS) materials has renewed interest in the possibilities for superconducting power apparatus offering real economic benefit, within power ratings typical of present system practice. Previously developed low-temperature superconductors (LTS) required cooling by liquid helium to about 4.2K, with advanced cryogenic technology that is expensive both in terms of cost and of refrigeration power expended per unit of heat power removed from the cryostat. The technology for the new materials, which may be based on liquid nitrogen (LN2) at temperatures up to about 78K, is simpler and cheaper, and the ratio of refrigeration power used to heat removed is reduced from over 1000 to about 25. There is significant activity around the world regarding applications of HTS materials to cables, motors, generators, fault current limiters, energy storage devices and transformers.

Properties of HTS Materials

The announcement in April 1986, by Muller and Bednorz (IBM) of superconductivity in the perovskite structure Lanthanum-Barium-Copper oxide at 30K, was an important step towards a wider application of superconductivity. This was followed by the discovery of Wu and co-workers in January 1988, of $Y_1Ba_2Cu_3O_{7\text{-}\delta}$ (YBCO or 123), with a transition temperature of 93K, bringing superconductivity above the boiling point of liquid nitrogen (77.4K at 1Atm), a cheap and widely available cryogen. There has since been much effort on the search for new materials, and the optimization of processes for production of thin films (<1μm), thick films (10-100μm), bulk materials, wires and tapes in single or multifilament composites. Many practical problems remain to be solved, but the potential for engineering application is clear.

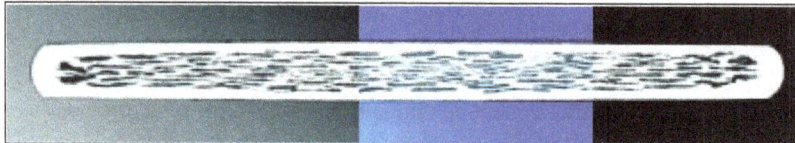

Multi-filament HTS tape.

Unlike LTS materials, HTS superconducting ceramics have highly anisotropic electronic structure, which causes critical current and critical field to have different values on two perpendicular planes. Randomly oriented polycrystalline HTS superconductors have low transport critical current, due not to intrinsic material properties but to misalignment between crystallites and weak superconducting regions at grain boundaries. Alignment or texturing of the material during the process of crystal growth improves properties and can be achieved by various means: temperature gradients, zone melting, two-dimensional configurations, etc.

The most promising known HTS materials for high current applications are given below, with their common numerical reference names in brackets:

Yttrium compounds (YBCO):

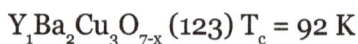

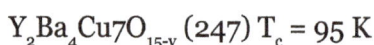

$Y_1Ba_2Cu_3O_{7\text{-}x}$ (123) T_c = 92 K

$Y_2Ba_4Cu7O_{15\text{-}y}$ (247) T_c = 95 K

Bismuth compounds (BISCCO):

$$Bi_2Sr_2Ca_1Cu_2Oy \qquad (2212)\ T_c = 80\ K$$

$$Bi_{2-x}Pb_xSr_2Ca_2Cu_3Oy\ (2223)\ T_c = 110\ K$$

Thallium compounds:

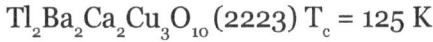

$$(TlPb)1Sr2Ca2Cu3O9\ (1223)\ \ T_c = 120\ K$$

$$Tl_2Ba_2Ca_2Cu_3O_{10}\ (2223)\ T_c = 125\ K$$

Mercury compounds:

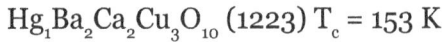

$$Hg_1Ba_2Ca_2Cu_3O_{10}\ (1223)\ T_c = 153\ K$$

Modeling and Simulation

HTS materials exhibit strong flux creep effects. When modelling power loss mechanism due to a flow of alternating current it is therefore not sufficient to use a critical state model, which has proved successful when dealing with low temperature superconductors, and it is necessary to consider the flux creep E–J characteristic. A flux creep region is well described by the Anderson-Kim model in which,

$$E = K_1 \sqrt{H} \sinh\left(K_2 \sqrt{H} J\right),$$

where K_1 and K_2 are parameters related to temperature and the properties of the material, or a simpler relationship suggested by Rhyner where,

$$E = E_c \left(J / J_c\right)^\alpha, \quad \rho = \rho_c \left(J / J_c\right)^{\alpha-1}.$$

The critical current density J_c corresponds to an electric field E_c of $100\mu Vm^{-1}$, and $\rho_c = E_c/J_c$. The power law contains the linear and critical state extremes ($\alpha = 1$ and $\alpha \to \infty$ respectively). It has been found that for practical HTS materials $\alpha \approx 20$ and thus the system is very non-linear.

Using the Rhyner model the governing field equation takes the following form in a two-dimensional space,

$$\frac{\partial^2 E}{\partial x^2} + \frac{\partial^2 E}{\partial y^2} = \mu_0 \frac{\partial}{\partial t}\left\{ \sigma_c \left|E\right|^{\frac{1}{\alpha}-1} E \right\},$$

where $E = E_z/E_c$ for brevity and $\sigma_c = J_c/E_c$. Using a rectangular space mesh $\Delta x \times \Delta y$, a Finite Difference scheme may be built which yields,

$$\left|E_{ij}^{(k+1)}\right|^{\frac{1}{\alpha}-1} E_{ij}^{(k+1)} = \left|E_{ij}^{(k)}\right|^{\frac{1}{\alpha}-1} E_{ij}^{(k)} + \Delta t \cdot C_{ij} = K_{ij},$$

$$E_{ij}^{(k+1)} = \left|K_{ij}\right|^{\alpha-1} K_{ij}.$$

where,

$$C =$$

$$\left\{\mu_0 \sigma_c \left(\Delta x\right)^2\right\}^{-1} \left\{\left(E_{i+1,j}^k + E_{i-1,j}^k\right) + R^2 \left(E_{i,j+1}^k + E_{i,j-1}^k\right) - 2\left(R^2 + 1\right)E_{i,j}^k\right\}$$

and R=Δx/Δy. The indices i and j denote the nodal addresses in the (x,y) space. The loss over a cycle can then be calculated using numerical integration as,

$$\int_0^T \int_V \left(J \cdot E\right) dv \, dt.$$

The stability requirements for the numerical solution are severe and in order to make the process as fast as possible the time steps are adjusted at each step using the criterion,

$$\Delta t \leq \frac{\sigma_c \mu_0 |E|^{\frac{1}{\alpha}-1} E_c^{1-\frac{1}{\alpha}}}{2\alpha \left(\frac{1}{\Delta x^2} + \frac{1}{\Delta y^2}\right)}.$$

Field penetration in HTS tapes can be modelled accurately as a highly non-linear diffusion process. The knowledge of AC losses is of paramount importance in the design of HTS power devices as these losses are released at low temperature. Figures show typical results of simulations.

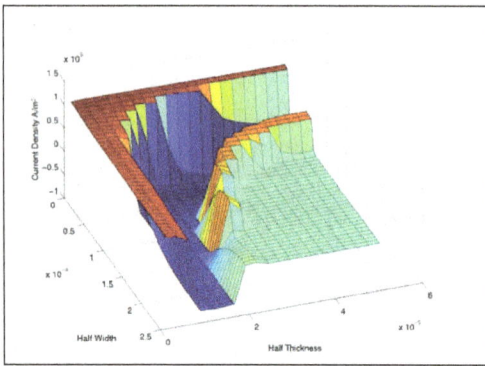

Current density distribution at the instant of a quarter of the maximum current.

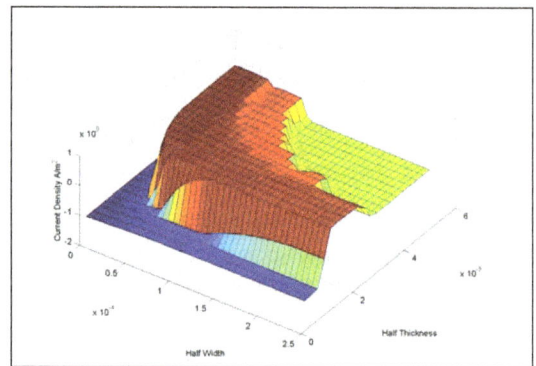

Current density distribution at the instant of max current.

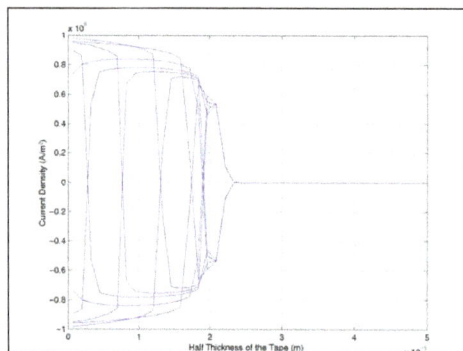

Current density profiles through the tape thickness.

The effects of non-linearity have been found to be much stronger than experienced in linear conductors and thus for larger currents (deeper field penetration) the 2D analysis is essential, despite a large aspect ratio of the tape. The analogy with the linear case only works well for smaller currents where a simpler 1D model is quite acceptable.

240Mva Grid Autotransformer

A design feasibility study was conducted for a 240MVA high temperature superconducting grid auto-transformer. The principal feature of the design is the removal of the copper windings and their replacement by HTS equivalents. These are only a fraction (less than 10%) of the bulk of the conventional windings. The inevitable result is windings of reduced mechanical strength, which will stand neither the radial bursting forces nor the axial compressive forces that occur during through-fault conditions without special strengthening structural features. The tap winding is kept outside the cryostat to avoid the heat which could otherwise flow into liquid nitrogen through a large number of connections. With the bulk of the ohmic losses (in common and series windings) removed, it is possible to cope with the core loss and remaining ohmic loss (in the tap windings) by forced gas cooling, leading to an oil-less design of transformer. This has a great advantage of reducing fire risk and environmental hazard from oil spillage. Furthermore, the need for an explosion-proof outer steel tank is removed – though some form of enclosure must be provided for weatherproofing and acoustic noise reduction.

The superconducting windings have small thermal mass and are cryogenically stable (i.e. capable of returning to normal operation after a period of abnormal heating without disconnecting and cooling down) over only a small range of temperature rise. In consequence, the HTS design has very little capability to recover from a through fault without disconnection, in contrast to a conventional transformer. This is the chief weakness of the HTS design; however, the HTS design can survive a through fault, though it subsequently needs disconnecting for a period, and there is also good overload capability.

The principal parameters of the suggested transformer design are listed below:

kVA: 240,000

Normal Volts: 400/132 kV

Tappings: 132 kV + 15% - 5% in 14 steps

Line current at normal volts: 346/1054 A Diagram No: Yyo Auto

Guaranteed reactance: 20%

Rated current densities:

series winding* = 39.1 A/mm² common winding* = 36.9 A/mm²

tap winding = 3.0 A/mm² (conventional).

(* average over composite conductor section, comprising both superconducting and matrix materials).

A short summary of the principal features of the HTS design is provided below:

- Core: Similar to a conventional design, but with leg length and window spans reduced and consequent saving in core weight and overall size.

- Tap change arrangements: Copper tap windings retained, adjacent to core leg, outside the cryostat, connected to a tap-changer, both of conventional design.

- Cooling of core and tap winding: Forced-convection gas cooling, probably nitrogen. Gas is forced through axial ducts to cool the core legs and tap windings. Top and bottom yokes fitted with fibreglass cowling to contain the fanned gas and direct it over yoke surfaces.

- Common and series windings: Constructed of rectangular section composite conductor comprising about 33% superconducting fibres and 67% matrix metal. The interterm insulating tape is applied to one face of the conductor. Individual tapes making up the conductor will be transposed in a normal way. Discs at the high-voltage ends of the series windings interleaved in pairs.

- Winding reinforcement: Outer diameter of series (high voltage) winding reinforced with fibreglass hoops (possibly pre-stressed) or continuous cylinder. Substantial inner cylinder with multiple spacing sticks supports inside diameter of common winding. Annular clamping plates top and bottom of complete in-cryostat winding assembly, pulled together by eight through-bolts or studs, all constructed of insulating material.

- Cryostats: One cryostat per leg, each cryostat comprising a vacuum vessel constructed of double-skinned fibreglass, with the vacuum continuously pumped. Cryostat pressure will be slightly in excess of 1 bar, containing nitrogen at 78 K. The intermediate-voltage leads pass through the top lid. High voltage lead passes centrally through cryostat wall.

- Fault and overload capability: For any substantial through fault, disconnection is required and a period of minutes is needed before reconnecting. Transformer can survive the most severe through fault for about 170 ms, within which time disconnection must be achieved. Internal faults sensed by in-built monitors or terminal voltage and current sensors, which initiate disconnection. Overload capability of 100% is expected.

- Housing: Oil-less design obviates need for a tank, which is replaced by a steel structure carrying load exerted by bushings. External housing required for weather proofing, gas-seal for the forced-convection nitrogen coolant and acoustic noise reduction.

Table summarises the total losses of the HTS transformer design and compares them with the corresponding figures for a conventional 'reference' design. Losses are expressed in percentage form with the total loss in conventional design taken as 100%. Table II shows all the significant global features, covering size and construction as well as performance.

Table: Loss analysis (total loss of conventional design = 100%).

	HTS	Conventional
Core loss	8	9
Clamp stray loss	5	5

Tank loss	-	7
Total copper loss	<1 (tap)	79
Refrigerator power	7	-
Gas-cooling fan loss	2	-
Estimated total loss	23	100

Table: Comparison of technical features.

Parameter	HTS	Conventional
Core length height	88.5	100
thickness	82.4	100
Window, height × width	100	100
	70 × 78.5	100 × 100
Core weight Winding weight	80	100
(common and series)	6.3	100
Tap winding weight	100	100
Cooling of core and tap winding	forced N2 gas	ONAN/OFAF
Cooling of common and series windings	liquid N2 (with refrigeration plant)	ONAN/OFAF
Guaranteed % reactance B in core, T J rated (average of C and S), rms, A/mm²	20	20
	1.67	1.67
Rated loss, total	38	2.83
	23	100
Overload capability	2 pu, many hours	1.3 pu, 6 hrs
Through fault capability, pu (+ doubling transient), recovery time without disconnection	2 pu, 64 ms	1.5 pu, 30 mins 5 pu, 3s
	166 ms	seconds (> 3)
Survival time at 5 pu		
(+ doubling transient)		

Table: Cost savings on continuous full-load.

Savings/(expenditure)	%
Saving on core plate	1
Saving on continuously transposed copper	7
Saving on copper losses, discount over 10 years	65
Cost of refrigeration plant	-21
First-cost equivalent expenditure on refrigerator drive power, discount over 10 years	-6

Cost of AC conductor, total of 7371 amp-kilometres	-10
Total equivalent first-cost saving	36

Table shows estimates of saving/expenditure components based on continuous operation in rated conditions. It is clear that an expenditure on extra equipment and materials is offset by the enormous value of the saved losses (taken over a 10-year period and discounted at 9.5% per year). However, because of the redundancy built into the system for security, the load factor of a grid transformer is remarkably low and may be taken as 0.23 average and 0.26 rms. For such load conditions a pessimistic estimate suggests that the total equivalent first-cost saving may now become a net increase of first-cost equivalent expenditure of about 20%.

On the other hand a common practice is to have two transformers fully rated normally connected in parallel. It is thus worth considering an arrangement of an HTS transformer normally connected, in parallel with a conventional transformer normally disconnected but capable of being switched on quickly when required (e.g. during through fault). Hence savings on the losses of the conventional transformer will be very significant.

Parallel Operation

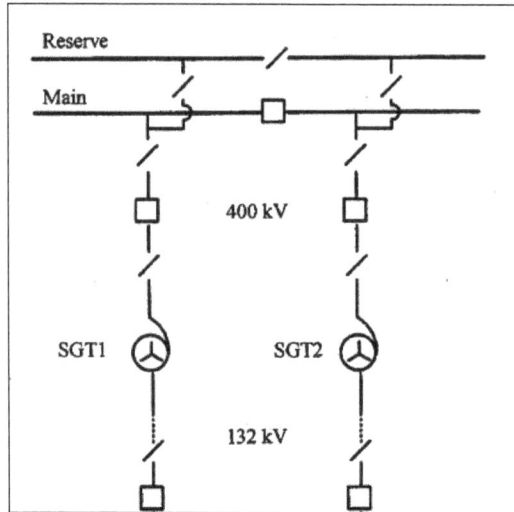

Conventional arrangement of parallel transformers.

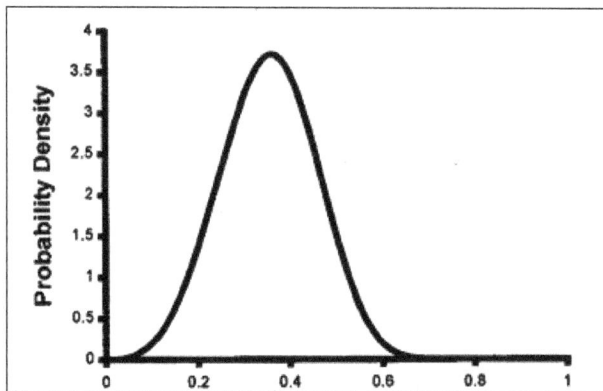

Probability density of load for a typical grid transformer.

Figure shows a typical arrangement of parallel connected transformers. Each transformer is usually rated to take the maximum load on its own, and because that maximum load rarely occurs, the load factor for a typical National Grid transformer in the UK is low, around 23%. A typical probability density function for such a transformer is shown in figure.

A suggested scenario is therefore to replace one of the conventional transformers with a HTS "equivalent". In normal operation the SGT2's breaker is open and the load passes through SGT1 (HTS). In the event of a through fault, SGT2's breaker is rapidly closed and SGT1's opened. This may require faster than usual breakers and associated protection but there is no reason to think that this may not be achieved within the 200ms required. Thus in this configuration it should hardly be necessary to ever use SGT2 (only during faults or maintenance outages) and so it may be possible to de-rate this transformer. The modified probability density distribution is shown in figure.

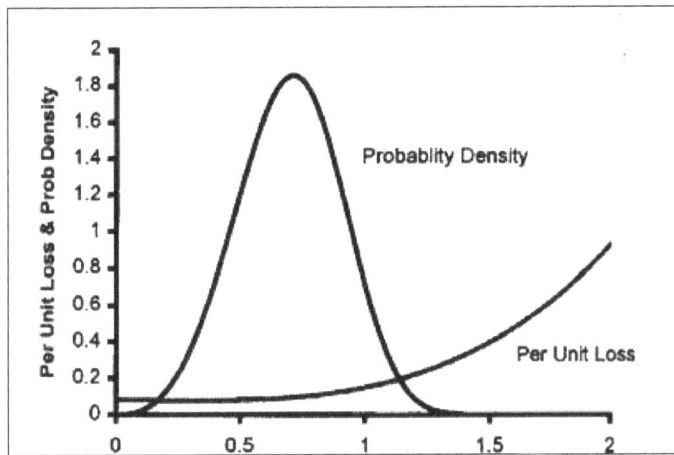

Load probability density and loss as a function of load for a HTS transformer in parallel with a (normally) unconnected conventional unit. The mean load is around 0.7 p.u.

Assuming a capital cost of a 240MVA transformer of £1.0M, per unit loss to be 600kW and cost of losses over the transformer life to be £3,000 / kW, we can assess the costs as follows:

Costs (£k)	Superconducting + conventional	2 × conventional
Transformer capital	1,000 + 1,230	2,000
Losses	0.105 × 600 × 3	0.426 × 600 × 3
Total	2,419	2,768

This simplistic analysis shows that the combination of HTS and conventional transformer is a lower cost option. Moreover, by inspecting figure, it is clearly possible to optimise this further by reducing the nominal rating of the HTS transformer thus reducing capital cost. It can be seen that the load would "hardly ever" go much above 1.5 p.u. so the HTS transformer could be of a smaller rating.

HTS Transformer Demonstrator

A small 10kVA HTS demonstrator transformer was designed, built and tested at Southampton University. In order to limit the material cost of this small single- phase unit (predominantly the cost of the superconducting tape), it was decided that the nominal rating at 78K should be 10kVA and only one winding, the secondary, should be superconducting; this also had the benefit of allowing direct comparison of performance between conventional and superconducting windings. The secondary current at this load is 40A. Since the large space required for thermal insulation between copper and superconducting windings increases the radial flux densities and leakage reactance, a 3 limb construction was used with the two windings on the centre limb. This arrangement minimises both of these problems, and also simplifies the design and construction. It is essential to reduce the radial component of leakage flux in the superconducting winding, as being perpendicular to the face of the tape it is most detrimental to its performance. For example, to carry the peak current of 9.5A per tape, the perpendicular component of peak leakage flux density must be less than 15mT, compared with 110mT for the parallel (axial) component. Field plots with and without the flux diverters. Flux diverters are placed close to the ends of the super- conducting winding in the cryostat and are constructed from low-loss materials to minimise the heat which must be removed from the cryostat.

Field plots with and without flux diverters.

The flux diverters are made from powdered iron epoxy composite. A small test ring of this material produced a relative permeability of 6, and the 50Hz iron loss at 78K with a peak density of 40mT is less than 1500W/m³. This loss is about 25% higher than that at room temperature due to the lower resistivity (the eddy-currents are resistance limited). A more expensive alternative, using flux diverters constructed of segments of ferrite with a relative permeability in excess of 100, was considered. However finite element analysis showed that, although the higher permeability did reduce the radial flux density in the end coils, circulating currents would be more difficult to control. The expected increase in the maximum capacity of the transformer is therefore very small and cannot justify the use of ferrite rings.

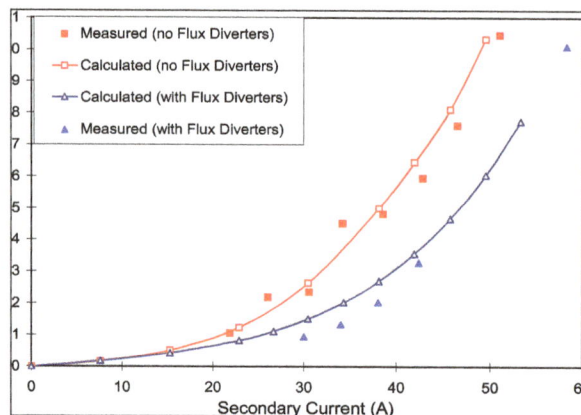

Measured/calculated losses with and without flux diverters.

Flux diverters have proved to be a very effective means of controlling local field distribution and reducing the winding losses. Figure demonstrates such losses as a function of secondary current. The graph also validates the numerical procedures developed for estimating AC loss in HTS windings. The most important observation resulting from these results is that the design requirements differ quite significantly from conventional (that is non-superconducting) transformers where the local shape of leakage field would not have mattered, whereas for the HTS winding a two-fold reduction of loss has been achieved. This emphasises the importance of very careful field modelling using finite element or similar software.

The BPSCCO-2223 tape of the HTS winding has 37 filaments in a silver matrix of outer dimensions 3.72×0.24mm. The nominal critical current in zero applied field at 78K is 20A, giving a bulk critical current density of 22.4A/mm^2. Each 28-turn disc consists of two tapes in each layer. There are nine discs and so 18 half-discs are available to form the six parallel paths required to carry the secondary current in the presence of the leakage field.

HTS winding of the 10kVA demonstrator.

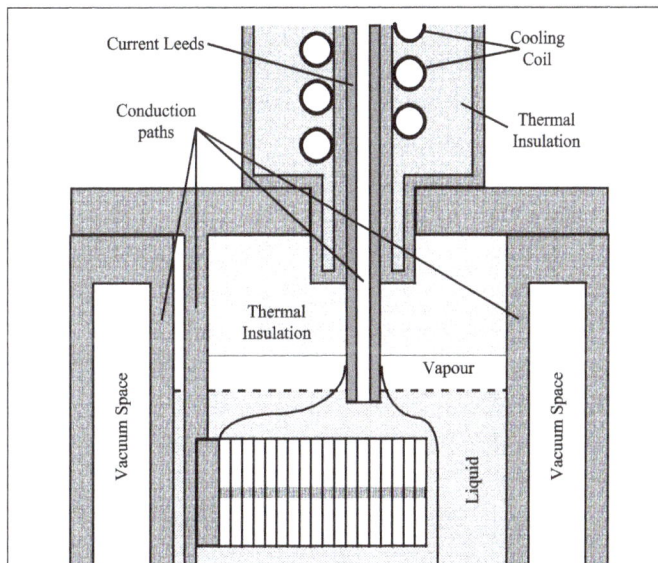

The cryostat and current leads.

The epoxy-filled fibre-glass cryostat consists of an inner annulus – which contains the HTS winding and liquid nitrogen – fitting inside the outer annulus. The cavity between the two halves is filled with superinsulation and evacuated to form the thermal insulation around the inner cryostat.

The whole assembly is threaded by the centre limb of the transformer laminated core. The complicated shape of the cryostat is necessary because it is not feasible to operate with a "cold" core in view of the loss of nitrogen that would be caused by the iron losses. Neither can the cryostat be made of electrically conducting material because it would act as a short-circuited turn.

Superconducting Super Motor and Generator

Structure and Features of the Super Motor/Generator

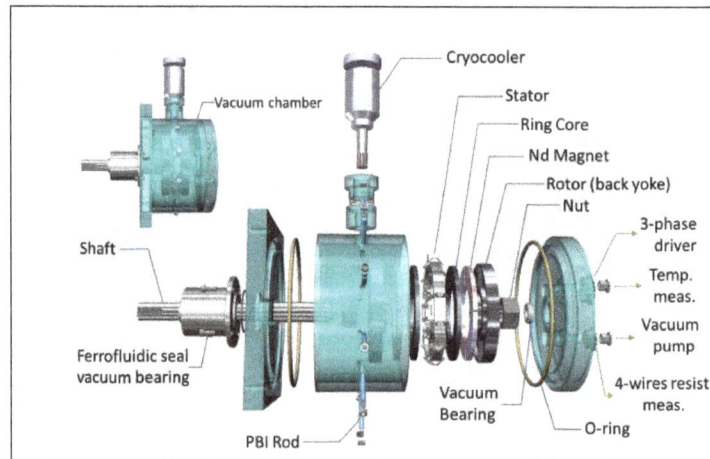

Schematic diagram of the prototype super motor/generator which is a synchronous four-pole machine.

The structure of the prototype super motor/generator is schematically illustrated in figure. The two rotors and the stator are placed in a single vacuum chamber, and the shaft is supported with ball bearings shielded with ferrofluid to maintain a high vacuum. Each of the rotors is composed of four Neodymium permanent magnets with a back yoke, that is, it has four poles, with a diameter of about 25 cm. The stator between the rotors is supported by 12 rods of polybenzimidazole (PBI) for thermal insulation. The stator itself is made of metal for good thermal conductivity, and contains two 10 mm thick toroidal cores wound from silicon iron tape, one on each side. The superconducting 2G-HTS wires are wound on the stator, and then impregnated with cryogenic epoxy resin. This stator is directly refrigerated below liquid nitrogen temperature by a compact cryocooler, and wrapped with superinsulation composed of 20 layers of polyester foil, aluminized on one side in 1 cm squares, interleaved with polyester spacer material. The total length of superconducting wire used is 100 m, with a tape width of 2 mm and a thickness of 0.1 mm. The thickness of the yttrium based superconductor itself is only 1 μm, while the tape base is made principally of copper and a nickel-tungsten nonmagnetic super alloy. The wide gap between the stator and rotors (for this prototype, 20 mm) easily contains the wire and insulation, permitting the wire length to be extended up to 1 km.

In this super motor/generator, the cores play two crucial roles. Firstly they efficiently apply the magnetic field to the superconducting wires to generate a Lorentz force, whose reaction causes torque, while avoiding applying magnetic fields to the metallic stator body, which would induce eddy currents. The flux from the north pole of the permanent magnets penetrates perpendicularly to the core, eventually leaving perpendicular to the edges of the core to reach the neighbouring south poles.

Secondly by making the mutual inductance almost equal to the self-inductance they enable the cancellation of the self-induction produced in the cores by the superconducting currents. This cancellation gives several remarkable features: the elimination of inductance enables a perfect power factor and high speed operation; prevention of magnetic saturation in the cores allows reduction of their weight; and ac losses in the superconductor due to the self-field are significantly reduced.

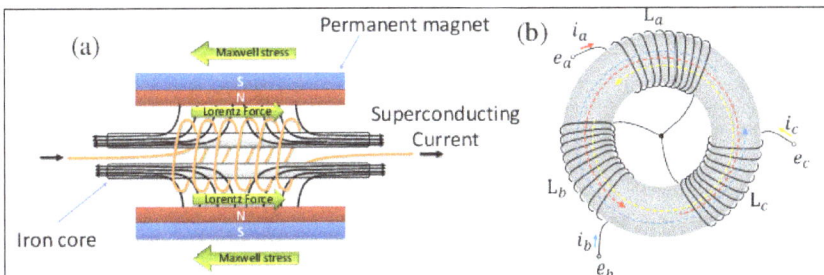

(a) Principle of torque generation in the super motor. (b) Schematic diagram of the armature coils. In the prototype each branch is wound as two separate coils of 90 turns each, on opposite sides of the stator and connected in series to give a six-pseudopole stator.

This cancellation is achieved by connecting the superconducting coils on the armature to each other, so that the total induction generated in the cores by the currents in any coils is almost cancelled out through mutual-inductances. Assuming that the three inductors are equivalent, that is they have same self-inductance, L_0, the mutual-inductance will be $L_0 - \Delta L$, differing from the selfinductance only due to flux leakage, which is greatly reduced by the cores. The circuit equations can be written in matrix form as,

$$
\begin{bmatrix} e_a \\ e_b \\ e_c \end{bmatrix} = \frac{d}{dt} \begin{bmatrix} L_0 & L_0 - \Delta L & L_0 - \Delta L \\ L_0 - \Delta L & L_0 & L_0 - \Delta L \\ L_0 - \Delta L & L_0 - \Delta L & L_0 \end{bmatrix} \begin{bmatrix} i_a \\ i_b \\ i_c \end{bmatrix} = \frac{d}{dt} \begin{bmatrix} \Delta L & 0 & 0 \\ 0 & \Delta L & 0 \\ 0 & 0 & \Delta L \end{bmatrix} \begin{bmatrix} i_a \\ i_b \\ i_c \end{bmatrix},
$$

where the second form can be obtained from the first in accordance with Kirchhoff's current law. The leakage inductance, ΔL, is expected to be very small relative to the self-inductance for toroidal cores, and so the input/output impedance is maintained at a low value in spite of the large number of turns. This feature allows large ac-currents to flow in the superconducting wires with little frequency dependence.

The Simulation

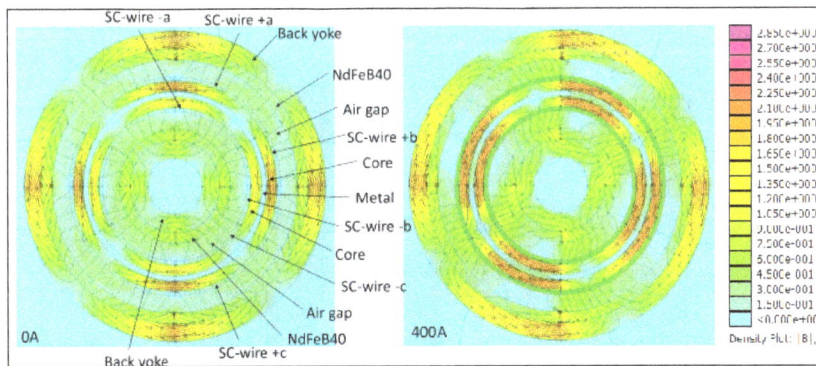

Magnetic flux distribution in the axial type super motor for currents of 0 and 400. A, where 90 turns/section. The colour scale runs from 0 to 3 T.

The magnetic field in the super motor has been calculated using finite element methods to estimate the torque and the power output which can be expected. In this simulation an equivalent cylindrical 2D model is used for convenience, instead of the actual 3D disk structure. Figure shows the magnetic fields for three-phase currents of amplitude 0, and 400 A, at zero degrees phase. It can be seen that the symmetry of magnetic field is broken as the current flow is increased. A torque acts on the permanent magnets as a reaction to the Lorentz force caused by the currents and the magnetic field, which can be represented as a Maxwell stress force.

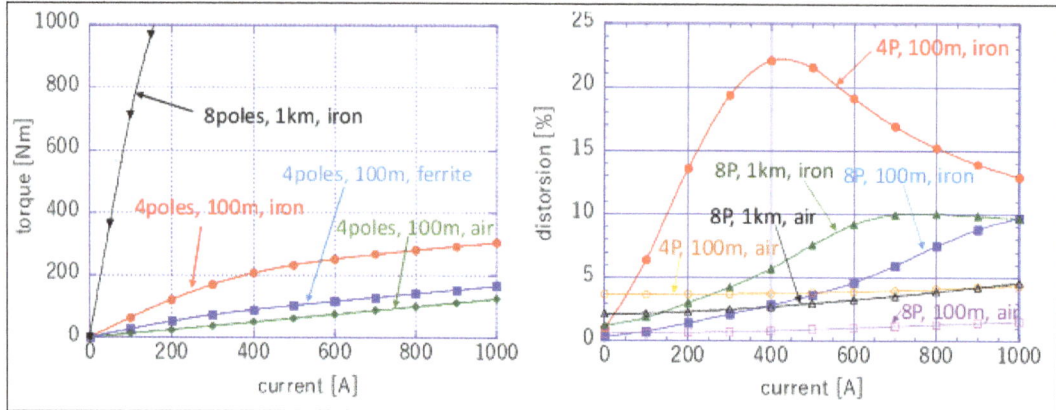

Calculated torque and torque distortion as a function of current for various machine models. The red curve (4 poles, 100 m, iron core) corresponds to the prototype.

Figure shows calculated torques and torque distortions as a function of current for three different core materials: silicon iron steel, ferrite, and air. Calculations for an eight-pole machine with 1km of superconducting wire are also shown in the figure. For air cores the torque is directly proportional to current, but the clear curvature for iron and ferrite cores arises from magnetic saturation effects. The calculated torque distortions are also primarily caused by partial magnetic saturation. Note that the distortions for eight-pole machines are much less than those for four-pole as a result of the reduced distance between poles.

Experimental Results

The prototype of the super motor/generator (left) and the induction motor (right) used for test driving.

A prototype was constructed and tested in our laboratory as shown in figure. The super motor/generator was connected to a conventional induction motor through a torque meter.

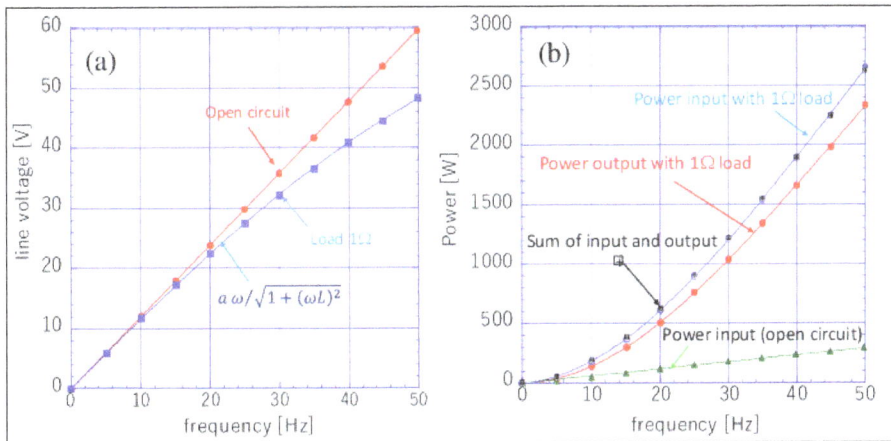

(a) Open circuit electromotive force and voltage drop across a 1 Ω load (Y-connection) as a function of frequency, which is twice the frequency of the driving motor as the generator has four rotor-poles. (b) Comparison of power output and input for open and 1 Ω loaded circuits. The sum (black crosses) of the open-circuit power input (green triangles) and the output on the 1 Ω loaded circuit (red circles) is almost equal to the power input on the 1 Ω loaded circuit (blue circles), indicating that ac losses from self-induction are negligible.

Power generation characteristics were measured with the super motor driven for a short time by a normal motor to eliminate temperature dependences (for example, the losses in the vacuum bearing depend strongly on the temperature of the ferrofluid and the superconducting armature). The open-circuit electromotive force and the voltage drops across an external 1 Ω load are shown in figure (a) as a function of frequency. By fitting these results to the function $a \omega R / \sqrt{R^2 + \omega^2 L^2}$, where R = 1 Ω is the external load, ω is angular frequency, and L is the residual inductance, the leakage inductance can be estimated at 2.2 mH. This measured result is four times larger than expected from simulations, indicating that the expected cancellation is not complete, perhaps because the superconductor wires are not wound neatly enough.

Figure (b) shows that the generated output power (calculated from the voltage across the 1 Ω external load) and the consumed input power (determined using the torque meter) as a function of driving frequency. The sum of the open-circuit input power input and the power output is consistent with the loaded input power, indicating that self-field ac losses in the superconducting wires are almost negligible, as expected due to the cancelation of self-induction.

Experimental torques, calculated from the generated output power and the driving frequency, which can be seen that torque is almost proportional to current from the good fitting with the function, as expected from the red curve in figure; we were unable to explore larger currents where significant curvature is expected to appear.

On open circuit most of the input power is consumed by friction in the ferrofluid vacuum bearing but ac-losses due to the rotating magnetic field are also included. These two components can be distinguished using a thermal method, as the ac-losses will be converted to heat in the stator, which is thermally isolated from the rest of the motor/generator assembly. Figure (a) shows how the temperature of the cold head changes as the driving speed was slowly changed between 300 rpm and zero, with the super generator on open circuit.

Thermal equilibrium temperatures were estimated as averages of the final temperatures from the falling and rising curves, and are plotted in figure (b). Assuming the Bean model with a critical current proportional to $T_c = 0$, the temperature below the critical temperature T_c, the ac losses are given by $Q(\theta) = a f (T_c - \theta)$ where f is the ac frequency in Hz. The cooling capacity of the cryocooler at constant input power is approximately linear near its limiting temperature, and can be described by $\theta_{eq}(f) - (fT_c + r\theta_0)/(f+r)$, where $r \equiv b/a$. This provides a good fit to the experimental data.

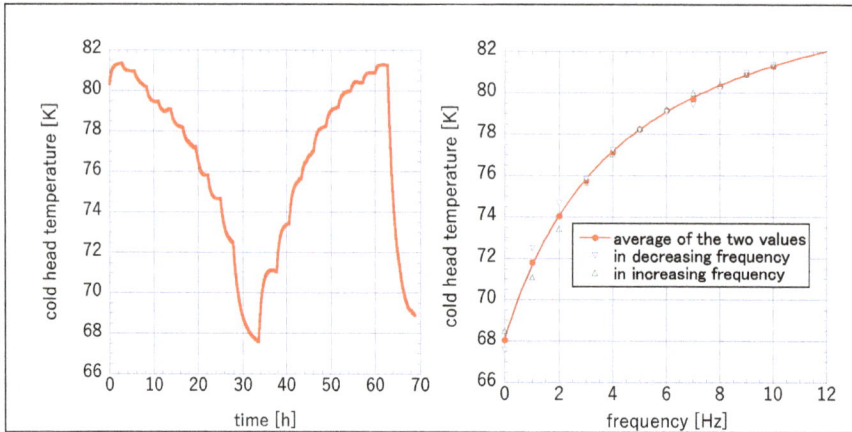

(a) Changes to the cold head temperature when the rotor is driven; the rotation speed is first decreased from 300 rpm to zero and then increased back to 300 rpm again. (b) The cold head temperature at thermal equilibrium as a function of driving frequency. The data is well described by $\theta_{eq}(f)=(fT_c+r\theta_0)/(f+r)$, with r=4.35 s^{-1}, T_c = 87 K, and θ0=68 K.

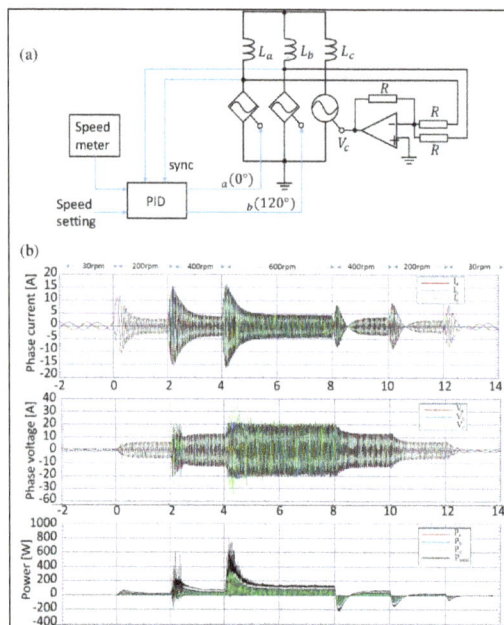

(a) Driving circuit for investigating motor operation. (b) The instantaneous currents, electromotive force, and power when driving with the circuit in the figure (a). The speed is first increased in steps of approximately 200 rpm and then decreased.

From this fit and the known value of b = 0.31 W/K we obtain a = 0.0713 W/Hz/K, and so Q(77K)=1.07 W/Hz. This value for the ac-losses is similar to a value of 1.24 W/Hz estimated from the experimental results of Amemiya assuming that the amplitude of the ac magnetic field is 0.276 T, estimated from the experimental torque.

Rotation experiments as a motor under light load were also performed with a driving circuit containing two current sources and one voltage source, where the electromotive force generated on the coil from the inductance L_c is cancelled by the voltage source. The instantaneous currents, voltages and powers for the three when the rotation speed is increased to 600 rpm in steps of about 200 rpm. From this it can be seen that the power is always positive during periods of steady driving, indicating that the currents and voltages are inphase and so the power factor is almost 100% in this current and speed range. When the rotation speed is decreased the consumed power is negative, indicating that the motor is acting as a generator, but during these short periods the power is always negative, so that the powe factor is once again 100%. Thus we conclude that the self-induced EMF is negligibly small relatively to the generated EMF in this experimental region. The speed range was limited by the available peak voltage (±20 V) of the current source, but the very high observed power factor assures effective operation over a much wider frequency range.

Silicon Superconducting Quantum Interference Device

The SQUIDs are composed of two weak links (Dayem bridges) acting as two Josephson junctions and are processed from a single layer of superconducting silicon. The superconducting film has been obtained by heavily doping the silicon with boron atoms using the Gas Immersion Laser Doping technique. This technique consists of shining laser pulses (here an XeCl 308 nm laser with a 25 ns pulse duration) at the surface of a silicon wafer on top of which molecules of BCl_3 have been previously chemisorbed. During a laser pulse, the silicon melts over a thickness and a time that depend on the energy density of the laser. In this melting time, the boron atoms diffuse very rapidly into the melted phase while the chloride atoms are expelled. At the end of the laser pulse, the boron atoms are incorporated in substitutional sites as the Si:B layer is epitaxially grown over silicon. Since the number of atoms that can be chemisorbed at the surface of silicon is self-limiting, the number of boron atoms introduced at each laser pulse is very reproducible and equals 1.2×10^{14} cm^{-2}. In order to increase the amount of dopants, the overall procedure (gas immersion and laser shot) is repeated (typically 10 to few hundred times) to reach active dopant concentrations up to 11% at with no boron aggregates. Such a high level of concentration, larger than the solubility limit of boron in silicon ($\simeq$ 1% at), can be achieved thanks to the rapid liquid/solid phase transition that quenches the boron atoms into the crystalline phase. In order to control the melting depth, the melting duration is monitored through the time resolved silicon reflectivity using a low power laser at 675 nm during each laser pulse. This procedure produces a very sharp doped undoped interface of only a few nm. In our set-up, the surface area covered by the laser beam is 2 mm × 2 mm. The study of the electronic transport properties down to very low temperature has revealed superconducting properties for a boron dose larger than 4×10^{15} cm^{-2} with a maximum of the critical temperature below which Si:B thin films show a zero resistance state of 0.7K for a boron concentration of 10%at and a thickness of 200 nm.

The superconducting silicon layer we have used for the present study was grown with 200 laser pulses and a melting time duration monitored to 47 ns. With these conditions, we obtained a 80 nm thick layer of boron doped silicon Si:B with a boron dose of $2.4 \times 10^{16} cm^{-2}$, which corresponds to a concentration of 5%at.

The fabrication of the SQUIDs starts with electron beam lithography to define the shapes (loops, weak links and contact electrodes) of the devices. Then, an aluminum hard mask (20 nm) is deposited by lift-off. More than 100 nm of the silicon not protected by the aluminium layer is etched away by reactive ion etching using fluoride gas and a highly anisotropic recipe. The aluminium mask is entirely removed afterwards by wet etching before the deposition of Ti/Au contact pads defined by optical lithography. Care has been taken to deoxydize the doped silicon before the deposition of the contact pads. A SEM image of a typical realization. The SQUIDs all have the same loop area and the dimensions of the weak links range from 80 nm to 200 nm in width and from 100 nm to 500 nm in length. In this letter we present the results obtained in a SQUID with weak links of 100 nm by 100 nm, but all the devices we have measured show magnetic flux modulation. We have also checked that the flux modulation in a SQUID loop can be reproduced using different silicon batches.

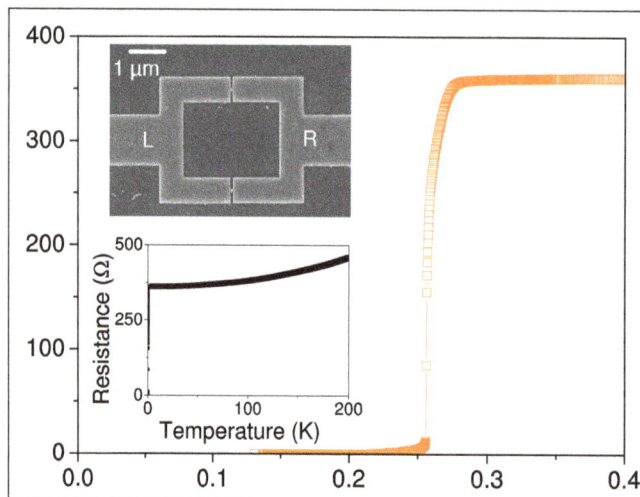

Resistive transition of a superconducting silicon SQUID. The SEM image shows the actual SQUID fabricated from a bare superconducting Si:B doped layer. It is composed of two weak links (Dayem bridge) with dimensions of 100 nm x 100 nm for the actual device reported here. Insert: Resistance of the device in the high temperature range showing a clear metallic behavior.

Figure shows the resistance of a SQUID as a function of temperature. The transition to the superconducting zero resistance state occurs at T_c = 260mK as expected for such a Si:B layer. The right insert of figure shows the resistance of the same device at high temperature up to 250K. The behavior is clearly metallic with a residual resistive ratio $R_{300K}/R_{10}K \simeq 1.4$.

At low temperature, a SQUID critical current I_c of several μA (corresponding to a current density of $\approx 10^4 A/cm^2$) is observed in the current-voltage I-V characteristics. Below 220mK the IV response is strongly hysteretic with a very small re-trapping current I_r. For current larger than the critical current, the resistance, defined as the slope of the I-V curve, is 360, which is exactly that of the device above the critical temperature. This means that the en- tire structure turns normal in the dissipative state. This behavior is due to a heating effect that propagates from the weak links to the rest of the device, as is usually the case for such devices. The temperature dependence of the critical current measured with no applied field is plotted. It shows a rather smooth variation with a critical current vanishing at the critical temperature.

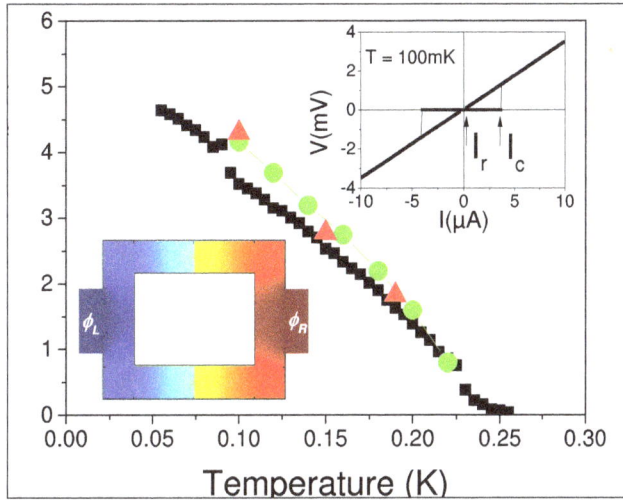

Temperature dependence of the measured critical current of the silicon SQUID (black square points). The green circles correspond to the maximum critical current obtained by measuring the magnetic flux modulation and the red triangles result from solving the Ginzburg-Landau equations. Left insert: color plot of the phase evolution of the superconducting wave function along the device for a phase coherence length of 40 nm and an applied phase fixed to ±3/4 at both ends of the device. Right insert: IV characteristics at 100mK showing a strong hysteresis when ramping the applied DC current. I_c and I_r show the critical current and the re-trapping current respectively.

For superconducting Josephson weak links, the amplitude of the critical current strongly depends on the length of the bridge compared to the superconducting Ginzburg-Landau (GL) coherence length ξ. In the dirty limit $\xi = \sqrt{\xi_0 l}$, where ξ_0 is the BCS (Bardeen-Cooper- Schrieffer) coherence length and l the elastic mean free path. For superconducting silicon at our doping level $l \approx 2-3$ nm and $\xi_0 \approx 1000$ nm, which gives the zero temperature phase coherence length $\xi_0 \approx 40 - 50$ nm. In the actual geometry, the ratio between the length of the weak link and the GL coherence length is $L/\xi(0) \simeq 2.5$. The critical current of Josephson bridge junctions can be obtained by solving the Ginzburg-Landau equations. Following the code developed by Hasselbach, we have numerically solved these equations in two dimensions taking into account the exact geometry of our device. This calculation is done by assuming a macroscopic superconducting wave function that can be written as $\psi(r) = f(r)\psi_\infty e^{i\phi(r)}$ where both f and ϕ depend on position but ψ_∞ is a constant at a given temperature. The values of f(r) are fixed to f = 1 at the left and right sides of the devices (position R and L in the SEM image) which means that the amplitude of the order parameter is constant and equals to its equilibrium value in the electrodes. The simulation is done with a certain phase ϕ_R and ϕ_L imposed at the left and right electrodes (usually $\phi_R = \phi_L$). We then obtain the amplitude and phase everywhere along the device. For the simulation, we have assumed a perfectly symmetrical case where the critical current in the upper and lower branches are equal. The left insert of figure shows a color plot of the local superconducting phase for a total phase difference $\Delta\phi = \phi_L - \phi_R \approx 3\pi/2$. The simulation shows that the main phase drop occurs at the junction bridges. Doing the same simulation for various phase differences, gives the current-phase relationship of the device and in turn, the magnetic field dependence of the critical current of the SQUID. The temperature dependence is obtained by taking into account the temperature dependence of the

GL phase coherence length that modifies the ratio $L/\xi(T)$ with $\xi(T) = \dfrac{\xi(0)}{\sqrt{1-T/T_c}}$. The results for

the critical current at three distinct temperatures is plotted in figure. They must be compared, not to the critical current obtained at no applied magnetic field, but to the maximum of the critical current flux modulation at different temperatures. Indeed, due to remanent and earth field contributions, zero applied field does not necessarily mean zero magnetic field. The agreement with experimental data is very good. The general temperature dependence obtained here shows a behavior very close to what has been obtained previously for similar values of $L/\xi(0)$.

One can also estimate the product $R_N I_{c,wl}$ where R_N is the normal state resistance of one bridge junction and $I_{c,wl}$ its critical current. The square resistance of the doped silicon film at low temperature can be extracted from the normal state resistance of the entire structure taking into account the geometry of the SQUID. Since the bridge is a square, the normal state resistance of one junction is therefore $R_N = R_\square \approx 14\Omega$, in agreement with the known resistivity of $R_N = R_\square \approx 14\Omega$ at such a doping level and thickness. Considering that the SQUID is symmetric, we find $R_N I_{c,wl} \approx 35\mu V$ with $I_{c,wl} = Ilc/2 = 2.5\mu A$. This value can be compared to the superconducting gap $\Delta/e \approx 40\mu V$ that we estimate from the critical temperature $T_c = 0.26K$ using the BCS relationship $\Delta = 1.76k_B T_c$. We then obtain $R_N I_{c,wl} \approx \Delta/e$ in agreement with existing calculations.

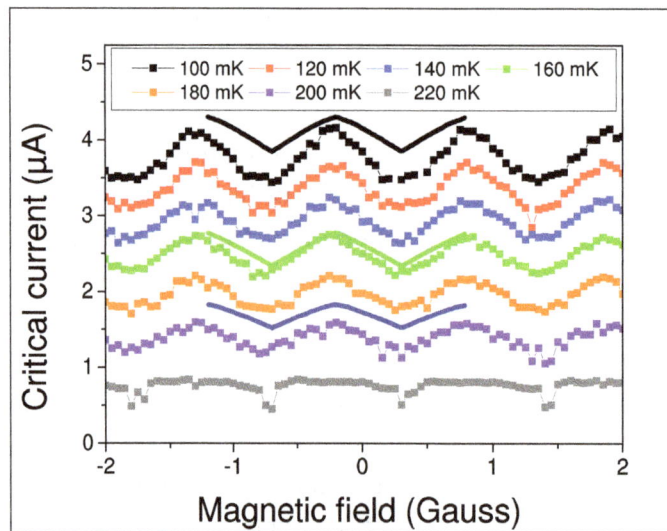

Magnetic field dependence of the critical current of the DC SQUID at various temperatures from 100mK (top) to 220mK (bottom). The measured modulation (scattered points) is compared to the flux dependence obtained from the GL equations in the actual geometry for three different values of the phase coherence lengths 40 nm, 60 nm and 80 nm to account for the temperature dependence (T ≃ 100mK – top, 150mK – middle and 190mK – bottom) (solid lines).

The dependence of the critical current with the magnetic field applied perpendicular to the device, for various temperatures. Each of these curves has been obtained by ramping the magnetic field from −2G to +2G in steps of 0.1G. The results show a very regular oscillation of the critical current as a function of the magnetic field with a period of 1G. This value corresponds to a one flux quantum $\Phi_0 = h/2e = 2 \times 10^{-15}$ Wb in a surface area of 20 μm², which is the surface of the SQUID loop 4 μm × 5 μm. The amplitude is roughly 10% of the maximum critical current. This flux (or field) modulation decreases when increasing the temperature and vanishes to zero when approaching the critical temperature. The solid lines in figure are the results of the critical current modulation obtained by numerically solving the GL equations. This has been done for three different values of the GL coherence lengths: 40 nm, 60 nm and 80 nm corresponding to 100mK, 150mK and 190mK approximatively. The position of the curve follows rather well the temperature dependence and the amplitude of the modulation is well reproduced too.

References

- Superconducting-wire: suptech.com, Retrieved 14 July, 2019

- Superconducting-magnet-technology-and-applications, superconductors-materials-properties-and-applications: intechopen.com, Retrieved 27 April, 2019

- Application-of-Superconductive-Equipment-In-Power-Industry-277022754: researchgate.net, Retrieved 18 June, 2019

- "Superconducting cable for power transmission application", Nexans – Superconducting Cable System (Hanover-Germany) – A Review

- Investigation of the Use of Superconductivity Drive to Make Moving Bodies Highly Efficient, SEI Technical Review, No. 168 (March 2006)

Permissions

Index